Interpretation of Organic Data

by

Roy-Keith Smith, PhD

Apichemical Consultants
9251 Highway 166
Winston, GA 30187

January, 2000

ISBN 1-890911-19-4
Printed in Canada
April, 2000

INTRODUCTION AND PREFACE

I have in other books[1] tried to present in considerable detail the requirements for acceptable analysis of environmental samples. I have divided these requirements into the categories of analytically valid and legally defensible, and spent about equal time writing on each. The term "evaluation" has been used to mean the analytical worth of the data and "validation" to describe the legal compliance side. However, in practice, what I have noticed in the environmental industry is that some persons who review data tend to spend a lot more time on the legal acceptability rather than the analytical utility of results. There are many reasons for this. It's easier to know and apply the dictates of "thou shall" and "thou shall not." Validation of results for these dictates tends to be more of a bookkeeping exercise, and I may have been as guilty as any other in drawing attention to this essential component of data acceptability.

The Environmental Protection Agency (EPA) is currently in the process of debating and adopting the Performance Based Measurement System (PBMS) as a paradigm for environmental analysis. In PBMS good science is stressed. This places an extra burden on persons reviewing data as they have to be able to recognize what is good science and what is not. The new edition of the *National Functional Guidelines for Organic Data Review* (EPA, 1999) has adopted this viewpoint with the key term "professional judgement" occurring on almost every page. My attempt in this book is to describe what is good science and what is not within the context of organic analysis under PBMS.

I have characterized this business as frequently frustrating but never boring. Every person I have ever met in this business has taught me something. The something may be positive or it may be negative, and in some cases it is accompanied by tense situations, but still my fund of knowledge grows. As always my thanks go out to my editor, Erik Roy at Genium, the owners of Analytical Services, Inc., Robert G. Owens, Jr, G. Wyn Jones, and Billy Paul Dyer, and all the colleagues, peers, and data reviewers with whom I interact in the environmental industry. I am not going to try to list them all, as someone might be slighted that they got overlooked. But if I have ever met you - and this includes all those who I have ever disagreed with - thanks.

R.-K. Smith 31 December, 1999

---

[1] Smith, R.-K., 1999. *Lectures on Wastewater Analysis and Interpretation*, Genium Publishing, Schenectady, NY; R.-K. Smith, 1999. *Handbook of Environmental Analysis, Fourth Edition*, Genium Publishing, Schenectady, NY; R.-K. Smith, 1995. *Water and Wastewater Laboratory Techniques*, Water Environment Federation, Alexandria, VA; W. Berger, H. McCarty, and R.-K. Smith, 1996. *Environmental Laboratory Data Evaluation*, Genium Publishing, Schenectady, NY.

TABLE OF CONTENTS

LIST OF FIGURES

LIST OF TABLES

CHAPTER 1

Separation and Sample Preparation

There are millions of organic compounds that have been characterized in the chemical literature, with still millions more that are waiting for attention. The biggest problem in organic analysis is isolating the single compound of interest from the rest of the organic soup.

Take for example the simple and very common compound hexane, C_6H_{14}. Hexane does not exist as a single unique compound. There are five structural isomers of hexane (Table 1-1), all with the same molecular formula. The mixture of hexanes available in a bottle for laboratory use are not homogeneous for the molecular weight 86 from the formula C_6H_{14}. There are additional compounds for each of these structural isomers that include the naturally occurring isotopes of hydrogen (deuterium) and carbon (carbon-13). Consider the simple substitution of one hydrogen atom with a deuterium. Among the isomers of the hexanes there are 18 unique compounds with just one deuterium. And for the substitution of a single carbon-13 atom there are 18 compounds.

Table 1-1. Structural isomers of hexane.

Compound	CAS No.	Structure
n-Hexane	110-54-3	
2-Methylpentane	107-83-5	
3-Methylpentane	96-14-0	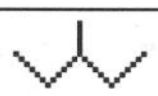
2,2-Dimethylbutane	75-83-2	
2, 3-Dimethylbutane	79-29-8	

In that bottle of hexanes, there are also contributions from closely related relatives of the hexanes. For the six-carbon relatives, there are the family of compounds with the formula C_6H_{12}, which includes molecules with a single double bond (13 compounds) or a single ring (seven compounds). These are also subject to having isotopic substitution, generating more than 100 unique compounds. If the five- and seven-carbon relatives are added, the number of compounds increases dramatically.

So when the task is to determine the presence of hexane in a sample, one of the objectives to be answered prior to any beginning of analysis, is how fine a discrimination is to be made in this determination. Is n-hexane, C_6H_{14}, Mol. Wt. 86 the target analyte, or will any compound that exhibits the same properties as generic hexane in the analytical system be accepted as hexane?

The above is a completely valid consideration that has to be applied to the design and implementation of organic analytical methods and to the evaluation of test results for any set of organic analytes. Analyzing a bottle of hexanes for specific isotopically substituted structural isomers is possible. However, the analysis of environmental samples is something of a Catch-22 in that the methods have been designed to be very robust so that they can handle all the complex crud associated with the sample, yet at the same time provide a measure of selectivity and specificity for isolation and identification of regulated compounds. It's akin to mixing the contents of the hexanes bottle with equal portions of gasoline and spent solvents, then being given the task of determining the isotopic distribution of the contents as well as the gasoline and spent solvents. It simply can not be done with a single analysis.

A. Analyte Isolation and Concentration

The first task in any analysis is to separate the analytes from the bulk of the sample. The traditional liquid extraction is the most common means employed. A portion of the sample is mixed with an organic solvent into which the analyte is preferentially partitioned. The idea of partitioning is key to the success of this procedure. No organic analyte is completely removed from a sample by a single washing with an organic solvent.

Partitioning is thermodynamically controlled. To achieve an equilibrium where, for example, 80% of the analyte molecules are in the organic solvent and 20% remain in the sample matrix, requires a certain amount of intimate contact time between the sample matrix, the analyte molecules, and the organic solvent. Gently sloshing together 1000 mL of aqueous sample and 60 mL of methylene chloride is not intimate contact nor is it efficient for achieving a thermodynamic equilibrium. The equilibrium is achieved by mixing the two phases together as thoroughly (violently) as possible for as long as possible, then allowing the phases to separate. Then repeat the procedure with additional portions of solvent. If equilibrium is established, then the first extraction recovers 80% of the analyte, the second extraction recovers an additional 16% (80% of the remaining 20% in the sample), and the third extraction recovers a further 3.2% for a total recovery of 99.2%. If equilibrium is not allowed to become established, the efficiency of the extraction procedure can be much lower.

All organic molecules have an affinity for the sample matrix to one degree or another, otherwise the organic molecules would not be present in the sample. The affinity may take the form of an actual absorption on the surface of sample particles or interaction with the water or other matrix molecules in a solvation phenomenon. Or it may take the form of an occlusion, where individual molecules of the analyte do not exhibit any attractive interaction with the substrate but are completely surrounded by substrate and have no means of exit from the cage.

It is necessary in an extraction process to overcome these interactions with the sample substrate. As a first attempt, and it must be stressed that, an attempt should be the operative concept, the sample is mixed with methylene chloride if the sample is aqueous, or, if the sample is solid, methylene chloride or a mixture of acetone and hexane. These are very powerful organic solvents and have great ability to partition analytes and other organic materials out of the sample. For some samples the great strength of these solvents actually works against the idea of separating the analytes from the matrix, particularly when the matrix contains non-analyte organic materials. When too much organic material ends up in the extraction solvent it is sometimes useful to switch to a solvent of lesser ability, such as hexane, pentane, or petroleum ether. In other cases, particularly when the sample is largely organic, such as an oil, the opposite tack can be useful, for instance, use of the polar solvent methanol to extract polar organic materials like the phenols from the matrix.

The organic acids (phenols) and organic bases (amines and pyridines) present other opportunities for sample isolation, where the pH of the sample can be adjusted to control the direction of partitioning. Acidification of the sample converts the organic bases into salts that preferentially partition into the water phase. Adjusting the pH of the sample to values >11 with a suitable base neutralizes the basic analytes and changes the direction of partitioning to the organic phase, while at the same time converting the organic acids to hydrophilic salts.

Selective isolation of analytes through pH adjustment has numerous applications. For instance, the isolation of phenolic compounds from ash and cement kiln clinkers is complicated by the high pH of the matrix. One solution is to use the basic pH as an aid to the isolation[1]. The solid sample is mixed with a large volume of water, 50 g to 1000 mL. The resulting water suspension exhibits a pH greater than 10 for most of these matrices. The slurry is extracted with methylene chloride to recover the neutral and basic organic analytes. Then the solids are allowed to settle from the water. The supernatent liquid is separated from the solids and acidified with sulfuric acid, then extracted with the organic solvent to recover the phenols.

The simple expedient of separate analysis of the acid and basic extractions, rather than combining the extracts into a single sample extract, often serves to reduce matrix interference to a manageable level. As illustrated later in Figures 1-4 and 1-5, the interference can be significantly reduced by this approach.

No two samples are identical. Each sample is a unique combination of matrix-analyte interactions. Adding surrogate compounds to each sample and then determining the recovery of the surrogates is used as a yardstick for gauging the success of the extraction procedure. The best surrogates are those that are most like the target analytes. Isotopically labeled versions of each of the target molecules are the ideal solution. Phenol-d_5 and 1,2-dichloroethane-d_4 are perfect surrogates for phenol and 1,2-dichloroethane. The isotopic dilution techniques are the ultimate extension of this idea, where the recovery of each labeled analyte is used as an internal correction for the efficiency of the analyte isolation procedure in the quantitation of each target compound. For most purposes, this level of confidence (and expense) in the analysis is not warranted.

As an alternative, a short list of representative surrogates is frequently used. The choice of the surrogates can limit or it can maximize the interpretational utility of the recoveries. For most analyses, it is possible to choose surrogates that will always generate excellent recoveries, regardless of the complexity of the sample. Choosing surrogates to produce "acceptable" QC is counterproductive. The idea behind surrogates is to obtain information about the weak points of the analytical procedure. In most organic analyses the sample preparation factors that need to be monitored through the surrogates are the chemical and physical behaviors of the target analytes. For samples that are to be analyzed by gas chromatography, the key physical behavior is the range of vapor pressures exhibited by the analytes. Important chemical behaviors include acidic and basic properties of the analytes, the range of polarity exhibited by the analytes, reactivity in derivatization procedures, and finally, the sensitivity of the analytes to decomposition caused by extremes of chemical or physical environment. Surrogates chosen to monitor any of these areas should ideally bracket the range of the property. For vapor pressure, a surrogate compound with a high vapor pressure used in conjunction with a surrogate with a low vapor pressure provides the maximum range of information. Examples of surrogates that generate these types of information are found in a variety of the EPA organic methods. However, it should be pointed out that very few individual methods specify surrogates that provide information on all these areas, let alone the idea of bracketing the property. Under the Performance Based

---

[1] Jackson, C., 1996. *Analysis of phenolic acid compounds in calcareous soils by SW-846 Method 8270,* 19th Annual EPA Conference on Analysis of Pollutants in the Environment, Norfolk VA 15-16 May, 1996.

Measurement System (PBMS) paradigm, additional surrogates beyond those specified in the relevant reference method can and should be used to extract the maximum amount of information about the sample.

Sample extract concentration is the one area in the isolation procedure that has the greatest potential for loss of analytes. The operative physical concept during concentration is vapor pressure, not boiling point. Boiling point is a discontinuous function: if the sample temperature is below the boiling point, it does not boil, if the sample is at the boiling point, it boils, and faster boiling is achieved by the addition of more heat. Boiling is uncontrolled: all the molecules are attempting to convert from liquid to gas at the same time. Equating sample concentration to boiling point leads to very erroneous conclusions about what is appropriate practice and what is not. Vapor pressure is a continuous function that relates to the rate of evaporation, and evaporation is what is desired during sample concentration. At low ambient temperatures vapor pressure is low, and evaporation is slow. At higher temperatures vapor pressure is higher and evaporation is faster. The conversion of a substance from a liquid state to a gaseous state requires heat, and the heat is generally extracted from the surroundings. On a hot day, if you wet your skin and then expose the skin to the air, your skin temperature is reduced due to the evaporation of the water, and you feel cooler. Obviously, the temperature of neither you nor your surroundings is at the boiling point of water, but the water still makes the transition from liquid to gaseous state.

These same processes are functioning during sample extract concentration. Conversion of solvent molecules from liquid state to gas is the desired objective, but the target analyte molecules are doing the same thing. It's just that, hopefully, the rate of conversion is slower for target analytes than it is for solvent. The proper technique is to control the rate of solvent evaporation while minimizing analyte loss. A completely successful sample extraction can be performed that is completely negated by an over-exuberant sample extract concentration.

Guidelines are normally presented in the methods, such that concentration of an extract below 1.00 mL is unacceptable because there is a substantial probability for loss of target analytes. As a rule this is useful, but it does not encompass all sample situations. There are samples that contain considerable amounts of co-extracted organic material, and even attempted reduction of the extract volume to 10.00 mL can result in complete loss of all potential target analytes.

The use of a surrogate that is more volatile than any of the target analytes is a very useful monitoring tool for concentration problems. Examples of surrogates that perform this function are tetrachloro-m-xylene (TCMX) in the chlorinated pesticides and PCB analyses, phenol-d_5 and 2-fluorophenol in the GC-MS analysis of semivolatile organics, and n-nonane in diesel fuel determinations (DRO). Low recoveries of these surrogates, in the presence of otherwise acceptable surrogate information, are almost always diagnostic for sample concentration problems. In this situation, recovery of the more volatile target analytes is regarded as similarly low, and data should be qualified as probably lower than the true values in the sample.

Acid-base properties of the surrogate must mimic those of the target analytes, otherwise the surrogate may be lost during sample manipulation. As an example, if the analytes are phenols, the surrogates must be phenols or other organic acids. A polyaromatic hydrocarbon such as phenanthrene is completely unsuited as a surrogate for phenols. A variety of isotopically labeled phenols are available for the mass spectral techniques, phenol-d_5 being the most common. EPA Method 8041 (determination of phenols) suggests use of 2,4-dibromophenol as surrogate for the non-mass spectral technique; however, any phenols that are not analytes, and are not present in the sample, can be used. Phenols are not the only organic acids that are subject to analysis. EPA Method 552 (determination of haloacetic acids) uses

3,5-dichlorobenzoic acid, 2,3-dichloropropionic acid, 2-bromopropionic acid, and 2,3-dibromopropionic acid as surrogates, depending on which version of the method is being followed.

When allowed choices in surrogates the analyst should make selections based on obtaining the maximum amount of information from the surrogates. In the case of the phenols, the compounds exhibit a range of ionization abilities, quantitatively expressed as the pK_a of the substance. Highly acidic phenols such as the nitrophenols have low pK_a values (2,4,6-trinitrophenol pK_a 0.42) and are more completely transferred to the water phase as salts than are the less easily ionized phenols (3-methylphenol pK_a 10.00). The nitrophenols are also less favorably partitioned into the organic solvent than the less acidic materials. In choosing surrogates for a phenols analysis, at least one surrogate should have a low pK_a while another has a high pK_a.

At present there are very few organic bases that are target analytes in environmental samples, as compared to the number of organic acids. Organic compounds that are classified as organic bases almost always contain nitrogen at the center of the ionization properties of the compound. However, not all nitrogen-containing compounds exhibit basic properties. True amines such as triethylamine and triethanol amine, are moderately strong bases. Aromatic amines (anilines) behave more like neutral compounds than as bases, and, although there are exceptions (4-nitroaniline, the benzidines, and the aniline dyes), are not partitioned into acidified water to any significant degree. Nitrogen-containing aromatic rings, such as pyridine, the quinolines, and other naturally occurring and synthetic alkaloids (caffeine and nicotine), are strong bases. Amides do not behave as bases. The strength of organic bases as bases can be expressed as either pK_a, or more commonly as pK_b, and these values can be used as a guide for selecting appropriate compounds as surrogates. Pyridine-d_5 has seen considerable use as a base surrogate, particularly in the analysis of wastewater effluents, but few other organic bases are employed for such purposes.

Organic analytes exhibit a wide range of polarity. Hexane and methanol are both organic compounds, but a mixture of the two solvents will form two distinct layers. There should be a polarity-matching between surrogates that are used in the analysis and the target analytes, particularly if any sample clean-up procedures are going to be used. One of the best tuned examples of this matching is the use of 1-chlorooctadecane in the Massachusetts extracted petroleum hydrocarbons (EPH) procedure. A silica gel column is used to separate the aliphatic hydrocarbons from the aromatic hydrocarbons. 1-Chlorooctadecane exhibits a polarity that is between that of the aliphatic and the aromatic hydrocarbons. The surrogate is monitored to insure that it ends up in the aliphatic fraction and not in the aromatic fraction. This assures that a complete separation of the aliphatics has been achieved.

In general, if separations are being used for sample clean-up, the best procedure is to use at least two surrogates. One should be similar in polarity to the target analytes, while the other more closely mirrors the interferents that are removed. This way both recovery and removal can be quantitatively monitored.

When derivatization procedures are used in an analysis, surrogates should be employed that can monitor the success of the reaction. The most common example is the use of a derivatization reaction to convert phenols to methyl ethers and organic acids to methyl esters. Diazomethane, trimethylsilyldiazomethane, boron trifluoride-methanol, and simply acidified methanol have been used for these conversions. Other derivatizations that are used include the conversion of phenols to pentafluorobenzylbromoethers (EPA Method 8041) and the conversion of biphenyl and chlorinated biphenyls to decachlorobiphenyl (EPA Method 508A). This last procedure is particularly finicky, and difluorobiphenyl is a satisfactory surrogate to use to demonstrate complete reaction. The surrogate product is octachlorodifluorobiphenyl, which is chromatographically distinct from decachlorobiphenyl.

Not all target analytes exhibit the same degree of reactivity in these procedures. A surrogate should be used that first and foremost generates information that the derivatization has, in fact, occurred successfully. Second, a surrogate should be chosen that represents the more difficult analytes to successfully derivatize. Using a surrogate that is instantaneously and quantitatively derivatized tells little about the more sluggish analytes. An example of such is the use of pentafluorophenol as a reaction monitor for the derivatization of pentachlorophenol, a compound that is difficult to derivatize in most procedures.

Hydrolysis reactions, such as those used in the isolation of chlorinated acid herbicides, are potential points for analyte loss. A surrogate compound that is added to the sample in a form that is difficult to hydrolyze can provide useful information about the success of the hydrolysis reaction. For the chlorinated acid herbicides, target analytes in the form of esters are the most difficult to recover, even though in agriculture the esters are a frequently applied form of the compounds. The use of a surrogate like an ester of 2,4-dibromophenoxyacetic acid can provide the needed information. The ester that is chosen can be simple like the methyl ester, or it can be complex and very representative, such as the wettable ester from propylene glycol butyl ether (CAS 1928-45-6).

The sensitivity of target analytes to chemical decomposition can be monitored. The chlorinated hydrocarbon pesticides and the nitrogen-phosphorus pesticides are very sensitive to sample extraction and clean-up procedures. Chemically labile surrogates are monitored to insure that the analytes have not been subjected to abusive conditions. Dibutylchlorendate (DBC), a surrogate used in the chlorinated pesticide procedures, is very sensitive to pH conditions during the sample processing and will easily hydrolyze when the pH exceeds the 5-9 range, giving low recoveries. Low recoveries of the surrogate then are related to low recoveries of any of the labile target analytes (endrin, the DDT family, the endosulfan family). An alternate use for this particular compound is to monitor the acid-permanganate clean-up that is used in the PCB analysis (EPA Method 8082). If the sample extract has been adequately treated during the clean-up there should be no trace of DBC in the chromatogram.

The nitrogen-phosphorus analysis (EPA Method 8141A, September, 1994 version) suggests use of tributyl- and triphenyl-phosphates as surrogates. These compounds are essentially inert to all the abusive conditions that could completely prevent any meaningful recovery of target analytes from the sample. To monitor these aspects of sample preparation, an additional labile surrogate should be added to the analysis.

Related to chemical decomposition is absorption and loss of the target analytes to the glassware and other equipment that is used during the processing. The chlorinated acid herbicides are famous for exhibiting low recoveries due to absorption to glass surfaces. Although it is difficult to choose a surrogate that is specifically oriented toward the absorption problem, the use of a hydrolysis surrogate that is wettable (esters that contain one or more hydroxyethyl functions) can serve a dual function.

B. Resolution

After the sample has been extracted and the extract been through clean-up, the next step is the separation of the individual components through gas chromatography (GC) or use of liquid chromatography (LC). The idea is to have all components in the sample extract presented as single, completely resolved peaks to the detector.

Resolution is the degree of separation between any two sequentially eluting peaks in an analysis. There are quantitative expressions for resolution. It can be expressed as the average of the widths of the peaks (W_a and W_B in Figure 1-2) divided into the difference in retention times (Figure 1-1):

$$\text{Resolution} = \frac{2[\text{RT B} - \text{RT A}]}{W_A + W_B}$$

However, few analysts will sit down and actually calculate resolution. Instead it's an eyeball evaluation. If two peaks are separate and distinct, they are said to be resolved. If the peaks partially overlap they are partially resolved, and if they substantially overlap, they are not resolved. See Figures 1-1 and 1-2.

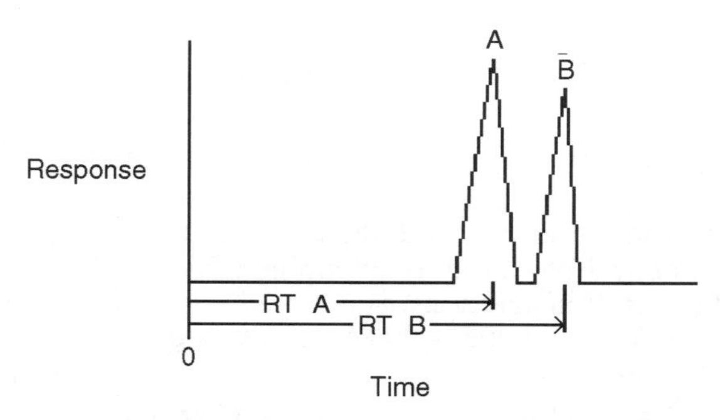

Figure 1-1. Retention times of two peaks in a generic chromatogram.

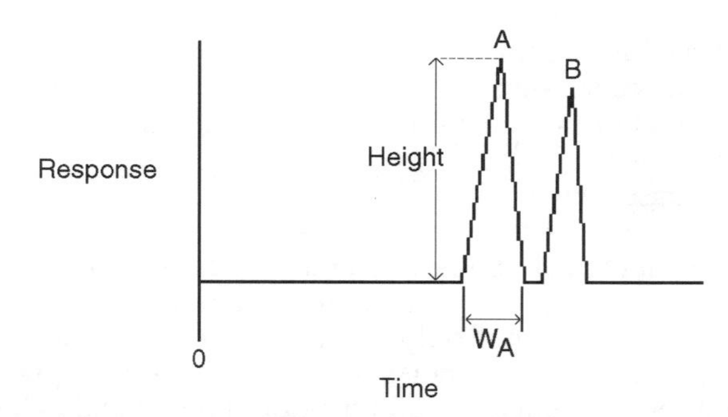

Figure 1-2. Characteristic properties of peaks in chromatography.

Peaks that overlap present problems in quantitation. The quantitation problems exist in the initial calibration of the instrument and in the analysis of samples.

Although the desired state is to have each and every component exiting the column as a single unique peak, real samples often contain co-extracted non-target compounds that co-elute with the calibrated analytes. Most times the analyst has to settle for completely isolated calibrated compounds. This can be difficult to achieve for 100% of the target analytes, particularly when some of the analytes are structural isomers of each other. Well known examples of representative analyte pairs are benzo(b)- and benzo(k)fluoranthene in the PAH and BN/A analysis, and 2-amino-4,6-dinitrotoluene and 4-amino-2,6-dinitrotoluene in the explosives residue determinations (Figure 1-3).

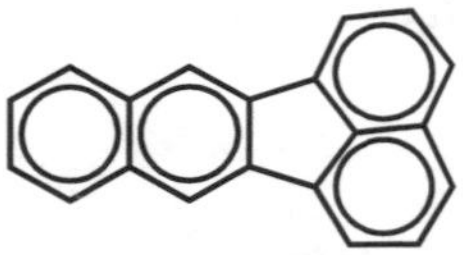

benzo(k)fluoranthene

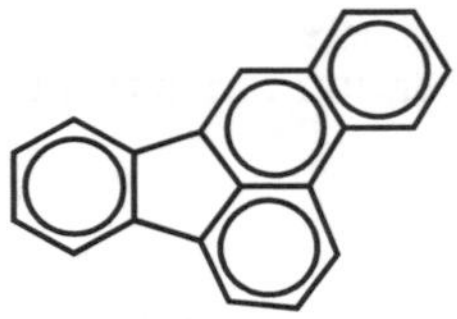

benzo(b)fluoranthene

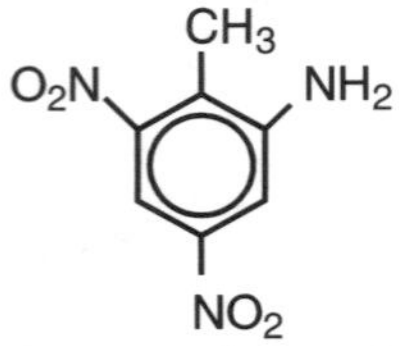

2-amino-4,6-dintrotoluene

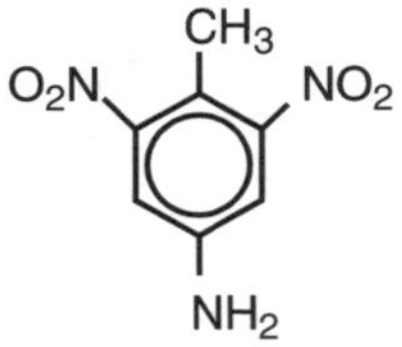

4-amino-2,6-dinitrotoluene

Figure 1-3. Examples of analyte pairs that are difficult to resolve.

Analysts have often resorted to the band-aid solution of using two or more sets of analytes to achieve accurate calibration of overlapping target analytes. One member of the pair is in one set of calibration solutions, and the other member is in a separate set of solutions. The overly optimistic hope is that both compounds do not appear in the same sample extract. The more correct approach is to closely examine and optimize the separation technique during the initial set-up of the analysis to avoid co-eluting target analytes.

There are a number of factors that will affect any given resolution problem. They are the:

- Mass of analytes on the column
- Temperature of the separation
- Chemical selectivity of the separatory column (stationary phase)
- Choice of mobile phase
- Rate of mobile phase flow.

The mass of analytes on the column is a corollary to the idea that for any given amount of liquid phase on column, there is a maximum amount of analyte that can interact with the liquid phase in a completely regular manner. The ideal operation of the separation is achieved when the molecules of the analyte will display a statistically normal distribution between being dissolved in the liquid phase and being in the vapor phase. As more molecules of the analyte are added to the system, they interact with each other, and the attraction to the liquid phase is as pairs or larger groupings of molecules rather than as single molecules. This changes the dynamics of the interaction with the liquid phase. Eventually the molecules' attraction for each other exceeds the interactive forces with the liquid phase and fronting or tailing occurs depending on the vapor pressure of the analyte-analyte interaction as compared to the analyte-liquid phase.

The solution is to inject no more analyte than that amount which will produce completely symmetrical peaks in the chromatogram. When completely symmetrical peaks are achieved, the mid-point of the peak is the ideal elution time (retention time) for the analyte, and the width of the peak is related to the amount of analyte on the column. As less analyte is placed on the column, the width of the peak decreases. When two analytes overlap, then one answer is to place lesser amounts of the two on the column and decrease the width of the individual peaks such that resolution increases, and the signals separate. This is easily achieved as a quick-fix by either decreasing the amount of sample extract solution injected into the instrument (a very poor choice due to non-linear discriminations in the injector port of the instrument) or diluting the sample extract with solvent and re-injecting the same volume of solution. The major consequence is that the lower limit of quantitation rises as the amount of analytes on-column decreases. More permanent fixes are to either use a longer column or use a column with a greater liquid phase thickness.

The temperature of the separation affects resolution. The vapor pressure (distribution profile) for the analyte-liquid phase is temperature dependent. At lower temperatures there is more interaction between the analyte and the liquid phase. When resolving two peaks, the lower temperature allows greater liquid phase discrimination over tiny differences between the analytes. The ideal situation is to keep the temperature of the analysis constant (isothermal operation), which allows the same level of interaction of the analytes with the liquid phase over the entire length of the column. A secondary effect of lower temperature is that the peaks become wider; however, the resolution increases faster than the broadening of the peaks.

Well, that's the ideal. In practical operation, the analyst is faced with a number of target compounds in the analysis that exhibit a range of vapor pressures. To complete the analysis in a reasonable amount of time, the column oven is temperature programmed, meaning that the temperature increases during the analysis. This allows analysis of a greater range of compounds in the GC run and also sharpens the peaks. However, it decreases the resolution as compared to an isothermal analysis. The rate of temperature increase affects resolution. Very fast temperature ramping severely degrades resolution. Slower rates of temperature increase give better resolution. It also makes the analysis time longer.

I once had the task of separating double bond positional isomers in straight chain hydrocarbons that ranged from C_{23} to C_{43} in length. The key compounds were those with double bonds in the 9-10 position as compared to those with the 10-11 double bond. A temperature program with a 1 °C/min increase from 200 °C to 320 °C achieved the needed resolution, but the analysis took 2 hours for each sample.[2] The point is that resolution of overlapping analytes can be achieved, it's just a matter of time.

The analyst needs to carefully consider the beginning of the GC analysis. The first signal out of the GC is the solvent front. Complete separation between the solvent front and the first eluting target compound must be achieved. Quantitation of early eluting peaks that are sitting on the downslope of the solvent peak is poor analytical practice and leads to a lack of precision in the determination. Many samples will exhibit early eluting co-extractables, and these can complicate identification and quantitation of the early surrogates and target compounds. A good practice is to have at least five minutes of separation between the last portion of the solvent and the first eluting calibrated compound. Again the trade-off is time.

The chemical selectivity of the column for the analyte is a key component in achieving resolution. It is also the most expensive component to adjust. Walter Jennings, the founder of J & W Scientific, a major supplier of capillary columns for GC, is reputed to have said that he could separate any two compounds with a dimethylsilicone liquid phase. This may or may not be true with regard to separations,

---

[2] Smith, R.-K., 1990. Chemotaxonomy of honey bees (*Apis mellifera* L.). Part 1: European and African workers. *Bee Science* 1(1):23-32; Smith, R.-K., 1991. Chemotaxonomy of honey bees (*Apis mellifera* L.). Part 2: Africanized workers. *Bee Science* 1(2):82-94; Smith, R.-K., M. Spivak, O.R. Taylor, Jr , C. Bennett, and M.L. Smith, 1992. Chemotaxonomy of honey bees (*Apis mellifera* L.). Part 3: Identification of Africanization in honey bee queens (Hymenoptera: Apidae). *Bee Science* 2(2): 93-105.

but the job can be made a lot simpler with a judicious choice of liquid phase. There are a lot of different liquid phases available for GC, and the situation is even more diverse on the LC side of the laboratory. If there is any one single item in regulatory methods that quickly becomes obsolete information, it is the column specifications in the method. Although one of the more current books on LC or GC technique may offer an idea of what is available, the better sources of information are the catalogs that are published by the suppliers. Most include sections on applications, and the analyst may find the exact solution in one of the applications. Time spent perusing column manufacturer's catalogs is not time wasted; it's time well spent. However, the best source of information is the technical support group of the supplier. These are the experts in matching columns with needed separations. When I used a GC regularly I was constantly on the phone to the various tech support centers of the major suppliers. Today you can contact the tech people by e-mail and do not have to sit on hold for extended periods of time. Do not limit yourself to a single supplier since no two companies offer exactly the same product line of columns. By asking the same question of a variety of people you improve your chances of getting the correct answer.

LC offers a wider selection of columns than GC does. Aside from the standard range of reverse phase octyl, octadecyl, and cyanopropyl columns, there are many quite unique columns whose like is not seen in the GC selection. These include the size exclusion, chiral, and immunochemical columns, one of which may be exactly suited for the needed separation. LC columns can also be connected in series to achieve particularly difficult determinations.

The mobile phase in GC is the carrier gas, and the gas selected has a bearing on the resolution. Nitrogen has very poor resolution ability, while helium or hydrogen are better choices. Hydrogen is actually the best carrier gas for resolution; however, it is reactive and may not be compatible with all sets of target analytes. There is an optimum flow rate for each carrier gas to achieve maximum resolution. As the temperature of the GC oven increases, the flow rate of the gas changes due to thermal expansion of the gas. Not so long ago (20 years) the analyst had to choose the temperature where he/she needed the maximum resolution and then set the carrier gas flow to optimize at that temperature. Most modern GC are equipped with constant flow devices that change the gas valve settings as the temperature in the oven changes, so changing flow rates are no longer a concern. Once the flow is optimized at one temperature it is optimized for all temperatures.

In LC there are as many or more choices of mobile phases as there are columns. Further the composition and flow rate of the mobile phase can be altered (gradients) during the analysis. The true sign of the expert LC operator is the ability to successfully choose the best combination of mobile and stationary phases to achieve needed resolutions.

C. Matrix Interferences

Many samples analyzed in the environmental laboratory display trivial-to-minor amounts of interfering substances, and the analysis is fairly simple. This is true for most wastewater samples and many groundwater samples, particularly those that are being collected as part of a long term monitoring program such as those associated with municipal wastewater treatment plants or landfills.

However, samples collected from sources that are heavily laden with organic matter present problems. A chromatogram of a wastewater sample extract from a chemical manufacturing plant is illustrated in Figure 1-4. The chromatogram exhibits a major amount of co-extracted interference that is localized in the 17- to 22- minute elution time window, and then a lot of minor interference that exists in all parts of the chromatogram. It should be obvious that identification and quantitation is going to be difficult for target compounds that would normally elute in the 17 to 22 minute range. The proper approaches to this type of problem are either to perform some sort of sample clean-up that effectively removes the interference from the sample extract or to use an alternate analytical instrument that is not affected by the interference. There are many clean-ups that are possible, although no one technique can be promoted as a universal solution to interference problems. Several useful techniques are discussed later.

Unfortunately the most common remedy selected by analysts is to dilute the sample extract by a factor of 5 to 10 or even up to 1000 in some cases depending on the amount of interference present, and re-analyze the diluted extract. The load on the detector of the instrument has been lessened by this technique, but the problem of sample interference is completely dodged. A significant side-effect of this technique is that when detection limits for target analytes are multiplied by the dilution factor a non-detect for an analyte can be at levels above regulatory limits, frequently viewed as evidence of a permit violation by regulatory officials.

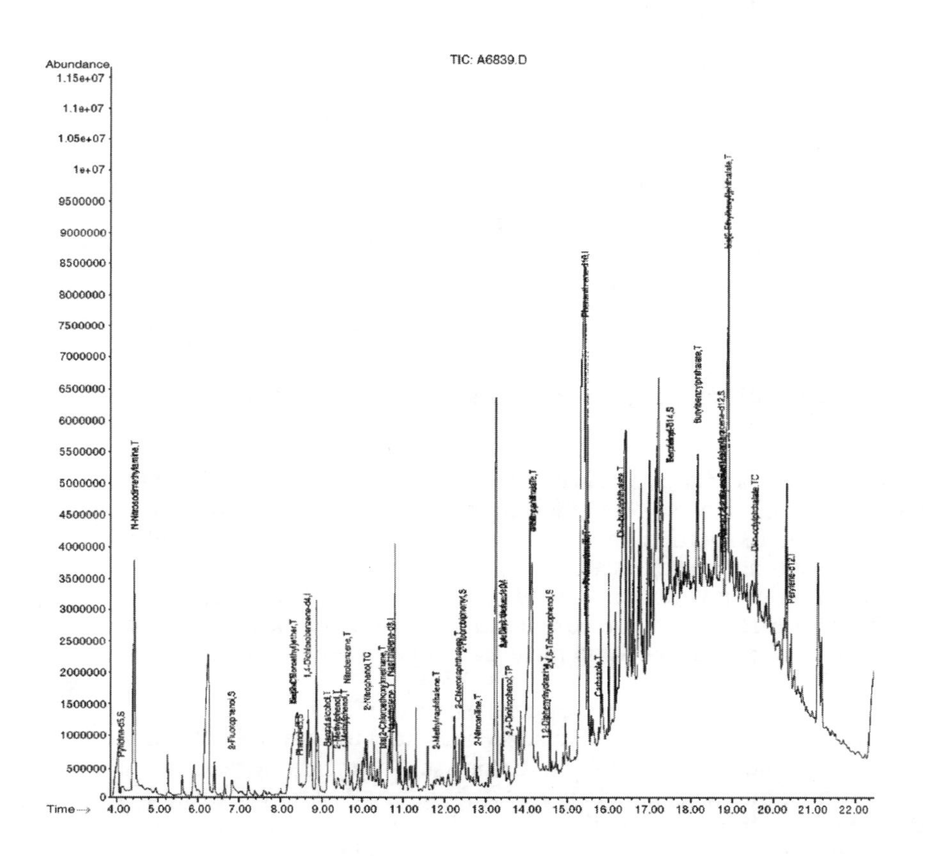

Figure 1-4. Acid extract chromatogram of a wastewater effluent sample.

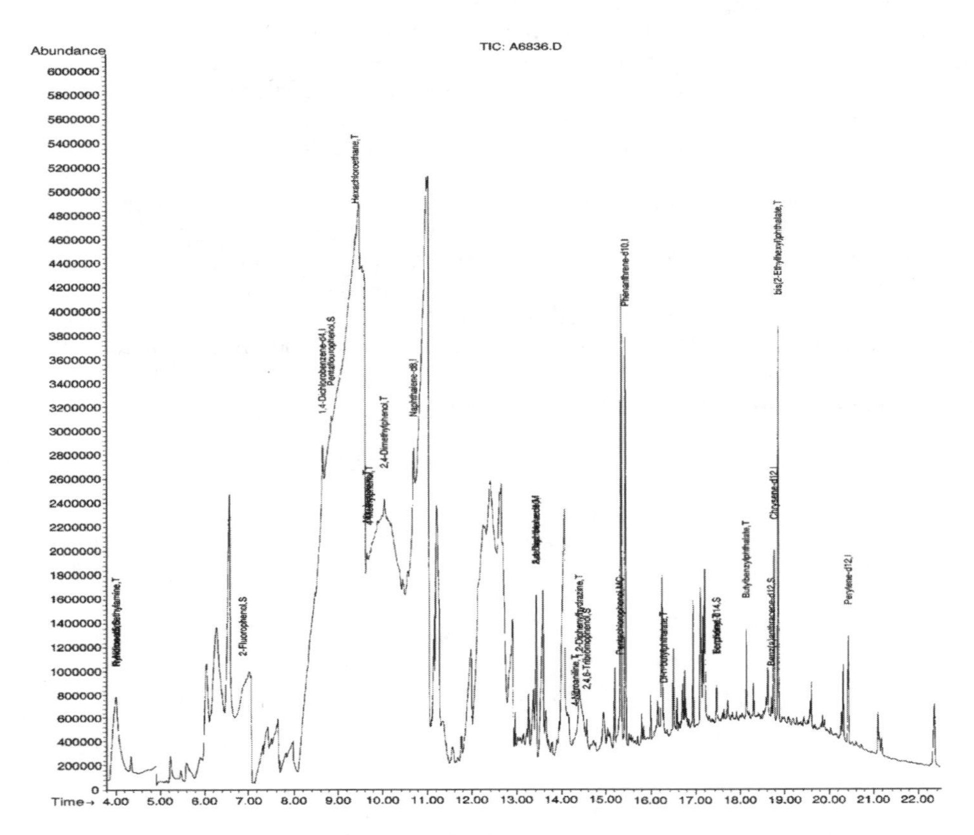

Figure 1-5. Basic extract chromatogram of a wastewater effluent sample.

Sample extract dilution and re-analysis has a very important but very limited role in the laboratory. The legitimate use occurs when a target analyte is found in the sample, but the amount in the portion analyzed is over the bounds of the calibration curve. The solution is to dilute the extract by a factor necessary to bring the amount of analyte in the analyzed portion within the bounds of the curve so that a proper quantitation can be performed. There are no scientifically valid reasons for performing sample extract dilutions as an alternative to the use of appropriate sample extraction and clean-up.

Some procedures dictate the use of a particular extraction procedure or clean-up prior to analysis that are viewed as minimum steps necessary to obtain reliable data. The use of Florisil® micro-column clean-up of chlorinated pesticide extracts is a required procedure under the Superfund Contract Laboratory Program (CLP). EPA Method 625 dictates the separate preparation and analysis of acid and base/neutral extracts of wastewater samples. The U.S. Army Corps of Engineers recommends use of gel-permeation chromatography for clean-up of soil extracts prior to analysis. Most of the regulatory methods of analysis, particularly the organic methods, contain flexibility so that the analyst can employ alternate methods of extraction and clean-up for problem samples, while still being in legal compliance with the method. The EPA has published in the methods manual for the solid waste program[3] a large selection of possible extraction and clean-up procedures (Table 1-2). I have discussed a number of these techniques and extensions, or modifications of them, in other publications.[4]

Table 1-2. Organic semivolatile sample preparation and clean-up procedures in SW-846.

3500B	Organic extraction and sample preparation
3510C	Separatory funnel liquid-liquid extraction
3520C	Continuous liquid-liquid extraction
3535	Solid phase extraction (SPE) (3535A in IVB)
3540C	Soxhlet extraction
3541	Automated soxhlet extraction
3542	Extraction of semivolatile analytes collected using modified Method 5 (Method 0010) sampling train
3545	Accelerated solvent extraction (ASE) (3545A in Update IVB)
3550B	Ultrasonic extraction
3560	Supercritical fluid extraction of total recoverable petroleum hydrocarbons (TRPH)
3561	Supercritical fluid extraction of polynuclear aromatic hydrocarbons
3562	Supercritical fluid extraction of PCB and organochlorine pesticides (Update IVB)
3580A	Waste dilution
3585	Waste dilution for volatile organics
3600C	Cleanup
3610B	Alumina cleanup
3611B	Alumina column cleanup and separation of petroleum wastes
3620B	Florisil cleanup
3630C	Silica gel cleanup
3640A	Gel-permeation cleanup
3650B	Acid-base partition cleanup
3660B	Sulfur cleanup
3665A	Sulfuric acid/permanganate cleanup

One clean-up technique in particular can be described as an example of the utility of these procedures. The analysis of polyaromatic hydrocarbons (PAH) is frequently needed on sites that are contaminated with petroleum fuels and oils. The PAH present significant risk to human health, and thus the required reporting limits for some of the compounds are normally quite low, on the order of 1 µg/L or lower for groundwater and 100 µg/kg or lower for soils. The presence of the other petroleum hydrocarbons serve as a significant chromatographic interference, as they are isolated along with the

[3] *Test Methods for Evaluating Solid Waste - Physical/Chemical Methods*, EPA/SW-846, 3rd Edition, 1986, Update 1, July, 1992, Updates II and IIa, 1994, Update III, 1996, and proposed Updates IVa and IVb, 1998. Available from the Internet at www.epa.gov/osw/methods or by subscription from the Government Printing Office.

[4] Smith, R.-K., 1999. *Handbook of Environmental Analysis, 4th Edition*, Genium Publishing, Schenectady, NY 1-800-243-6486.

PAH in the solvent extraction. Direct analysis by GC-FID or GC-MS of the extract without clean-up generates the familiar "hump-o-gram", from which the analyst will try to pick out the individual target PAH frequently resulting in detection limits above 1000 µg/L for water and 33000 µg/kg for soils, useless values for ascertaining remediation success. One possible remedy is to use a GC-PID for the extract analysis. The PID (photo-ionization detector) is significantly more sensitive to PAH target analytes than the FID (flame ionization detector) or MS (mass spectrometer), while at the same time being less sensitive to saturated petroleum hydrocarbons. This works for some samples; however, the capability of the PID can be exceeded in samples with heavy petroleum loadings.

The solution is to remove the petroleum hydrocarbon interference from the sample extract, and it is very easily done.[5] The methylene chloride extraction solvent is exchanged for hexane[6], then the sample is passed through a silica gel cartridge. The saturated hydrocarbons pass through the cartridge, while the PAH are retained. The PAH are then selectively eluted with either methylene chloride or ether. Analysis of the eluant by GC-FID or GC-MS gives a chromatogram uncomplicated by the saturated petroleum hydrocarbons, allowing PAH determination at very low detection levels, even when the original sample is soaked with oil.

The use of any sample clean-up in an analysis is always accompanied by the possibility of analyte loss. Procedures that depend upon polarity interactions between the eluting solvent, a solid phase absorbent, and the target analytes to achieve selective isolations, are particularly prone to having the desired compounds ending up in the wrong fraction. Sources of these errors include mistakes in the preparation of the eluting solvent, use of the wrong or a deactivated absorbent, and the presence of traces of polar solvents in the sample solution. Intense attention to detail and procedure are required for successful use of sample clean-up, yet even the most careful technician will occasionally botch the procedure. These fears may be a significant reason why most laboratories do not regularly use sample clean-up procedures.

Use of appropriate quality control procedures can go far toward building confidence about the sample processing. The aspects of the procedure that need to be examined include:

- Lot-wise suitability of the materials to achieve the clean-up
- Introduction of laboratory contamination
- Batch-wise success of the clean-up
- Success of the procedure on the individual sample.

Each new lot of materials, including solvent and solid phase absorbent cartridges must be checked to verify that they can achieve the desired result. Normally, processing a standard mix of the target analytes and a representative matrix interference with each new lot of materials, and measuring the recovery of the analytes and the removal of the interference is sufficient to verify the suitability of the materials. Problems that will be discovered through use of this quality control include manufacturer's pre-packed absorbent mini-columns containing the wrong or a

---

[5] This procedure is adapted from the EPH (extracted petroleum hydrocarbon) method for petroleum fractionation from the Massachusetts Department of Environmental Protection. www.state.ma.us/dep/bwsc/vph_eph.htm

[6] Remember that a single addition of hexane to the extract and concentration, leaves traces of methylene chloride in the sample. At least three cycles of addition and concentration are needed to completely remove the lower boiling solvent.

deactivated absorbent, and bottles of solvent that are contaminated with more polar substances.

The introduction of laboratory contamination is a significant concern in sample clean-up. Blanks are the most common quality control used to monitor introduced contamination. A common finding is the presence of phthalates extracted from some plastics that are used for construction of the mini-columns. Other contaminants that have been found include petroleum hydrocarbons in the C_{12} to C_{25} range extracted from the solid absorbent. Solvents can also be contaminated. Blanks should be processed through the sample clean-up each day that the procedure is used. The blank may be specific for the sample clean-up processing, rather than using the normal extraction batch blank; however, in most situations the batch blank will suffice. The blank is examined for the introduction of both target analytes and general contamination.

Batch-wise success of the clean-up can be determined through processing the batch laboratory control sample through the sample clean-up. The recovery of the target analytes is examined along with the introduction of any extraneous materials from the process. Although it may seem that this later function is simply duplicating that of the blank, in fact it is not. The laboratory control sample contains target analytes and exhibits a different polarity from that of the blank. The presence of the target analytes in the laboratory control can serve to displace column contaminants that are unaffected by passage of the blank. The laboratory control also serves as an additional check on random contamination.

Finally, it is desirable to check the success of the clean-up on every sample that is processed. Surrogates are normally used in organic analysis as an individual sample check. Surrogate compounds that are designed and used to monitor sample extraction may not be the most applicable monitors of a clean-up procedure. Take the example of the PAH clean-up described above. If the normal suite of six surrogates used in the Method 8270 analysis are subjected to the silica gel clean-up, four of the surrogates (phenol-d_5, 2-fluorophenol, 1,3,5-tribromophenol, and nitrobenzene-d_5) will probably not be recovered unless a much more polar solvent mixture for the final elution is used. Two surrogates should be recovered, 2-fluorobiphenyl and terphenyl-d_{14}; however, they fail to give information about the success of saturated hydrocarbon removal efficiency. For these reasons, a recovery surrogate, 1-chlorooctadecane is added to the sample extract prior to clean-up. The 1-chlorooctadecane should be completely absent from the PAH fraction, as it is eluted from the column with the saturated hydrocarbons. The recovery standards suggested in the MADEP EPH method, 2-fluorobiphenyl and 2-bromonaphthalene, are used to assure that all the lighter PAH, such as naphthalene and the methylnaphthalenes are retained in the aromatic fraction.

Another example of the use of a recovery surrogate is found in the sulfuric acid clean-up (EPA Method 3660 or 3665) of extracts for PCB analysis (EPA Method 8082). The clean-up procedure is to treat the hexane extract of the sample with concentrated sulfuric acid. In practice, most technicians will add the acid to the hexane extract and then allow the mixture to sit passively for several minutes. At most the technician may shake the mixture once, then let it sit. The only area of actual contact between the acid solution and the extract is at the solvent interface. This is a very ineffective situation. The mixture of the acid and the extract must be vigorously mixed, preferably with a small magnetic stirbar, to achieve a suitable level of contact in the clean-up process. To check that suitable contact has occurred, the addition of an acid-labile surrogate (dibutyl chlorendate, or 4-chloro-3-nitrobenzotrifluoride) to the sample or the sample extract is performed. Again the disappearance of the surrogate is indicative that adequate cleanup has been achieved.

CHAPTER 2

Detection

The term "chromatography" as applied to separation techniques, is literally translated from the Greek roots as "color picture." It stems from the original experiments where components of dyes were separated with filter paper as the stationary phase and water or other solvents as the mobile phase. The various components of the dye formed concentric colored circles as the solvent advanced through the filter paper. Most compounds that are the target of analysis in the environmental laboratory are colorless, and some alternate means of converting the chemical information present in the eluting peak is used to obtain information. In general these information conversion devices are called detectors, and there is slew of them available for attachment to gas chromatographs (GC) and liquid chromatographs (LC).

The most commonly employed detectors for GC in the environmental laboratory are the flame ionization detector (FID), the electrolytic conductivity detector (ELCD or Hall detector), the electron capture detector (ECD), the nitrogen-phosphorus detector (NPD), the photo-ionization detector (PID) and the mass spectrometer (MS). On the LC side the most common detectors are the ultraviolet detector (UV, used in both a fixed wavelength and a scanning mode), the fluorescence detector, and the mass spectrometer. In LC configurations where an ion-exchange column is used, a conductivity detector is frequently found. Less frequently encountered are the scanning infrared detector (IR), flame photometric detector (FPD), and atomic emission detector (AED) for GC. There are many other types of detectors that have been developed ranging from an oxygen-selective GC detector[1] to the non-discriminating refractive index detector used with LC and the thermal conductivity detector (TCD) in GC, however they have found little application in the commercial laboratories.

Detectors are characterized by the properties of sensitivity and selectivity. Selectivity is the response of the detector to a limited range of compounds that exhibit a common physical or chemical characteristic. The FID is essentially non-selective, meaning it will generate a response to almost any compound that can elute from the GC. The shared chemical characteristic of the FID-detected compounds is that they will burn in a hydrogen flame to form positive charged ions. Compounds that create a response in an ECD all are capable of preventing electrons from traveling between the ^{63}Ni source and the collector inside the detector.

The selectivity can be increased in a detector through the use of a post-column derivatization. The derivatization uses a chemical process to convert eluted analytes from one chemical form to another, which is then subjected to detection. Post-column derivatization is commonly employed in LC. One of the more common examples is the post-column derivatization of carbamate pesticides with sodium hydroxide, mercaptoethanol and phthalaldehyde (Figure 2-1) as presented in EPA Methods 531.1 and 8318. The isoquinoline that is the product of the reaction is detected using fluorescence-UV measurement. The ELCD is an example of a post-column derivatization used in GC.

---

[1] Kishi, H., H. Arimoto, and T. Fujii, 1998. Analysis of alcohols and phenols with a newly designed gas chromatographic detector. Anal. Chem. 70(16):3488-3492.

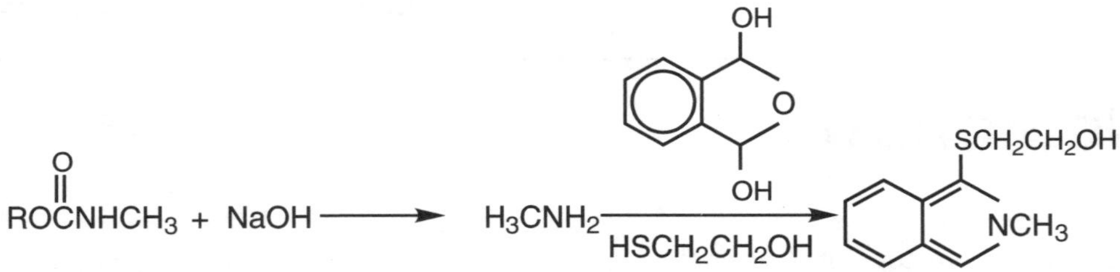

$ROC(=O)NHCH_3 + NaOH \longrightarrow H_3CNH_2$ (phthalaldehyde, OH, O, OH; $HSCH_2CH_2OH$) $\longrightarrow$ (SCH_2CH_2OH, NCH_3)

Figure 2-1. Post-column derivatization of carbamate pesticides.

The downside to post-column derivatization is that specialized equipment is required. The detector unit with the post-column derivatization is most frequently packaged as a single accessory for the GC or LC, and its use is limited to the single specific application. Most laboratories try to purchase versatile instruments rather than instruments with a restricted use.

Sensitivity is related to the lowest number of molecules that will generate a measurable response from the detector. It is a function of the efficiency of the interaction process between the compound and the physics of the detector.

Detectors can also be characterized by the results of the interaction between the compound and the detector. In most cases the molecules of the substance are irreversibly changed by the detection process. This is termed destructive detection, and examples are the FID, the ELCD, and mass spectrometry. A few detection techniques leave the molecules unchanged. These are termed non-destructive. Examples of non-destructive detectors are the PID, UV detection, and ECD. After the molecules of an eluting peak have passed through a non-destructive detector they can be further subjected to another detection process. EPA Method 8021B has the effluent from the GC column pass first through a PID and then on to an ELCD for destructive detection.

The detection of a peak eluting from a GC or an LC is the first step in a process that may lead to identification of a target analyte. There are two types of information derived from a detector. The first is the time of the response as measured from the moment of injection of the sample. This is called the retention time. All detectors are capable of generating this item of information.

The second type of information is the output of the detector as it responds to a chemical or physical property of the eluting peak. It is this type of information where the most confusion occurs. Most detectors are very general in response, meaning they respond to a large variety of compounds. Only a few detectors are selective, meaning that they respond to a property exhibited by only a limited set of substances. The most desirable, but virtually unattainable, type of detector is specific.[2] Yet many persons make the mistake of thinking that a chromatographic analysis is specific.

The root of the confusion lies in not separating the detector characteristics of selectivity and sensitivity. Just because an ECD is very sensitive, i.e. will generate a very large signal on a per molecule basis, to compounds that contain chlorine or bromine atoms, does not mean that it will not respond to other types of molecules. And, in fact, an ECD will generate a respectable response from molecules that contain nitro or carboxylate functions. Further, if a large amount of hydrocarbon material is introduced into the ECD, it will block the movement of electrons from the source to the collector, even though hydrocarbon molecules are very inefficient absorbers of

[2] Persson, B.-A., J. Vessman, and R.D. McDowell, 1998. Is your method specific or just selective? LC•GC, 16(6):556-560, June, 1998.

electrons. Evidence for this is the large hump-o-gram that is present in the ECD trace from analysis of polychlorinated biphenyls (PCB) in oils.

The output of most detectors is a continuous signal that is generated from measurement of a property of the compound under a single set of conditions. Most detectors fall in this category, such as the FID, ECD, PID, ELCD, NPD, and single wave-length UV monitoring. A plot of the signal against the time elapsed from moment of injection of the sample into the instrument generates the familiar chromatogram (Figures 1-4 and 1-5) with axes of time and response. The chromatogram is a two-dimension representation of the analysis, and detectors that are limited to generation of this picture are termed two-dimensional detectors.

Other detectors such as the scanning IR, scanning UV, and mass spectrometer, modulate the response signal through variation of a parameter (wavelength or mass/charge) during the course of obtaining information from the column effluent. This process generates an extra dimension to the detector output. In addition to the two-dimensional chromatogram, there is a spectrum (IR, UV, or mass) for each segment of time during the analysis. The duration of the time segment depends upon the cycle rate of the detector and the data handling capabilities of the attached computer. Most modern instruments will obtain several spectra per second of operation. These type of analyses are termed three-dimensional.

These three-dimensional detectors, as a result of the time required for cycling the data collection process, are generally less sensitive than a two-dimensional detector.[3] Sensitivity is increased in these instruments by decreasing the scan range and thus decreasing the cycle time, allowing collection of more data. The ultimate in sensitivity for these instruments is obtained by collecting data at only one wavelength or mass/charge unit for the duration of the analysis. This effectively converts the instrument from a three-dimensional mode to a two-dimensional mode of operation; however, the increase in sensitivity can be dramatic, as much as a 1000-fold or greater.

As a general comparison of the two types of detectors, the two-dimensional detectors produce good quantitation results and good sensitivity, but identification information of limited reliability. The three-dimensional detectors produce high quality identification information; however, they suffer with regard to sensitivity and quantitation precision.

A. TCD

The thermal conductivity detector is a non-selective detector. It measures the difference between the thermal conductivity of the plain carrier gas and that of the carrier gas mixed with substances exiting the column. It is non-destructive; however, the eluting peaks from the column are considerably diluted by extra carrier gas as they exit the detector. The threshold for peak identification is very high.

B. FID

The flame ionization detector is a non-selective detector. It measures the number of positively charged ions that are created from the hydrogen flame combustion of substances exiting the column. Ions of the type CH^+, CH_2^+, CH_3^+ from the burning of compounds containing C-H bonds are the most common detected; however, any

---

[3] One exception to this is the photo-diode array UV detector that can collect a continuous signal from each unit in the array. Each diode generates a two-dimensional signal, and collectively they produce a three-dimensional analytical result. However for the general case, the sensitivity is decreased.

substance that can generate a positive ion will elicit a response. The FID is a destructive detector. Compounds that do not burn, such as poly-halogenated substances, give poor response. Compounds that contain few C-H bonds such as formic acid HCO_2H give poor response. The FID has a very low background signal when no burnable materials are in the flame, and it is extraordinarily rugged, recovering very quickly to baseline after introduction of massive overloads of material. The FID generates linear calibrations for certain compounds up to a 5-decades range, meaning factors of 1 to over 10,000 in concentration. A schematic for an FID is presented in Figure 2-2.

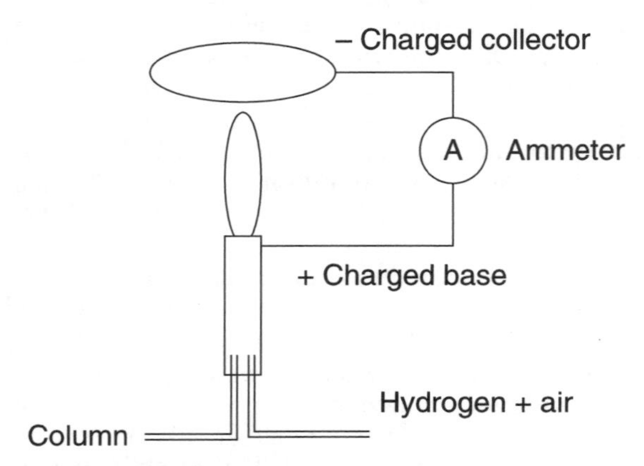

Figure 2-2. Schematic of a FID.

C. NPD

The nitrogen-phosphorus detector is a selective detector for compounds containing nitrogen or phosphorus atoms. It has a very similar construction to the FID, with the exception that a ceramic bead coated with a rubidium salt is positioned over the orifice of the jet. Hydrogen and air are mixed with the effluent from the column. An electrical current heats the bead, creating a plasma on its surface from the hydrogen. The plasma combusts the substances exiting the column and generates positive ions from nitrogen and phosphorus atoms in the compound. The NPD is a destructive detector, but unlike the FID, it is quite finicky and temperamental. It exhibits a low background standing signal, and is more sensitive than the FID due to the selectivity of the ionization. The calibration range for the instrument is quite limited, on the order of 2-3 decades for most analytes, with non-linear calibrations being commonly required. The NPD cannot tolerate overloads since they can de-activate the bead. The use of chlorinated solvents with the NPD will drastically shorten the life of the bead.

D. ECD

The electron capture detector is a somewhat selective detector. The detector consists of a source of electrons, normally a β-particle emitter such as ^{63}Ni, an ionization gas (argon or argon-methane) to multiply and moderate the electron flux, a positively charged collector, and electronics to allow counting of current. The background standing current is quite high. As substances are introduced into the detector that are capable of blocking the flow of electrons from the source, the signal decreases. The ECD is a very sensitive detector, responding to very low levels of materials. As a detector it is not very rugged, being very slow to return to background.

The usable calibration ranges are short, at the most a factor of less than 100 in concentration being common. Calibrations are frequently non-linear, particularly at the lower concentrations. An overload of material introduced into the ECD saturates the detector and can completely disable it. Materials that have a high electron capture cross-section will create the largest signal from the ECD. The ECD is non-destructive; however, the dilution of the analytes by the rapid flow of assessory gas makes further detection very difficult.

E. ELCD

The electrolytic conductivity detector is a very selective detector for substances that contain chlorine, bromine, or iodine atoms. Substances exiting the column are mixed with hydrogen gas, then passed through a nickel reaction tube that is heated to 900 °C. The halogens in the compound are converted to their respective hydrogen halides, HCl, HBr, or HI, thus the ELCD is a destructive detector. The gas stream from the reaction tube is mixed into a flowing stream of n-propanol. The n-propanol stream flows through a conductivity detector, which measures the increase in conductivity due to dissolved hydrogen halides. The background conductivity of n-propanol is very low, and the addition of a hydrogen halide increases the measured value dramatically. When operating, the detector is subject to few false-positive interferences; however, it is very high-maintenance. The ELCD has a limited calibration range.

F. PID

The photo-ionization detector is a somewhat selective detector for substances that are capable of becoming ionized by UV radiation. The UV light source can be changed to higher or lower maximum wavelength, which affects the range of compounds that will become ionized; however, the most common source is the 10-electron-volt lamp. UV light from this lamp will ionize aromatic compounds, substances that contain double bonds, and substances that contain oxygen atoms, such as methyl-*t*-butyl ether. The detector is non-destructive and is frequently used in series with an ELCD or an ECD. It is quite rugged, returning very rapidly to baseline after an overload of material is introduced into the device, and it is very low maintenance. Calibrations are linear and of 3-5 decade range, frequently exceeding the usable range of the halogen selective detectors that are used in series with it. A schematic of the PID is presented in Figure 2-3.

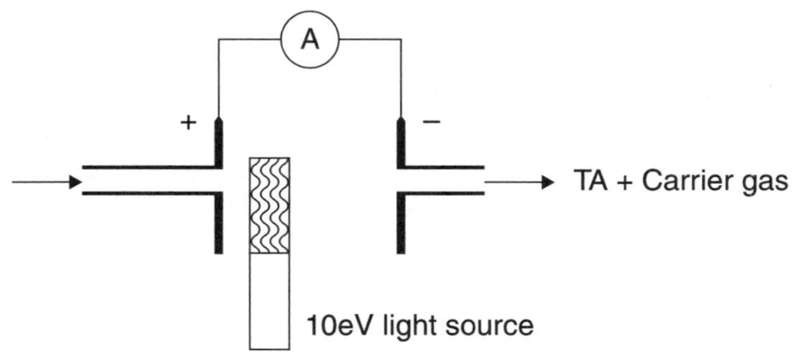

Figure 2-3. Schematic of a PID.

G. AED

The atomic emission detector is a selective destructive detector for a variety of elements. The trace for each element creates a two dimensional chromatogram. However, if more than one element is monitored, the AED can generate spectral-like three-dimensional results. The column effluent is introduced into a helium plasma, where the analytes are destroyed and the consituent atoms thermally excited and ionized. As the ionized atoms migrate to the outer, cooler portions of the plasma, they combine with electrons and energetically relax to lower excited levels, emitting photons of light at characteristic frequencies. Photodetectors are used to record the characteristic emitted light from the plasma. The AED is fairly rugged, and returns to baseline rapidly after analyte overloads. The detector can be set-up to monitor several elements at the same time, and if the analyte has a unique collection of elements in it, the AED can be quite selective. The AED has reasonable sensitivity, and a linear calibration for most analytes in the 3 decade range.

H. FPD

The flame photometric detector is a highly selective destructive detector for either phosphorus or sulfur, depending on configuration. A hydrogen flame is used to burn the sample. Sulfur in the sample is converted into excited S_2 molecules while phosphorus atoms generate excited HPO molecules. When either of these species migrate into the cooler portions of the flame they relax and emit light of a characteristic frequency. A photodetector equipped with a sulfur- or phosphorus-specific filter is positioned above the flame to record the emitted light. Calibrations are non-linear and of short range. Other limitations include interference by non-specific light generated from samples with high organic loadings.

I. Refractive Index

Refractive index measurements were one of the first detectors used with commercial LC instruments. The detectors are rugged and non-destructive, but not terribly sensitive. The detection mechanism is based on the refractive index of a mobile phase changing as other materials are dissolved into it. The detector cannot reliably be used with solvents of changing composition, such as is the case with gradient elution programs.

J. UV

Ultraviolet detectors can be configured in either the single wavelength two-dimensional or in a scanning three-dimensional mode. Either mode is non-destructive. The most common wavelength for the two-dimensional detection is 254 nm, although other set-points are sometimes used. Calibrations are generally quite linear, Lambert-Beers law being applicable. The range of calibration is restricted, primarily due to limited analyte capacity on the LC column.

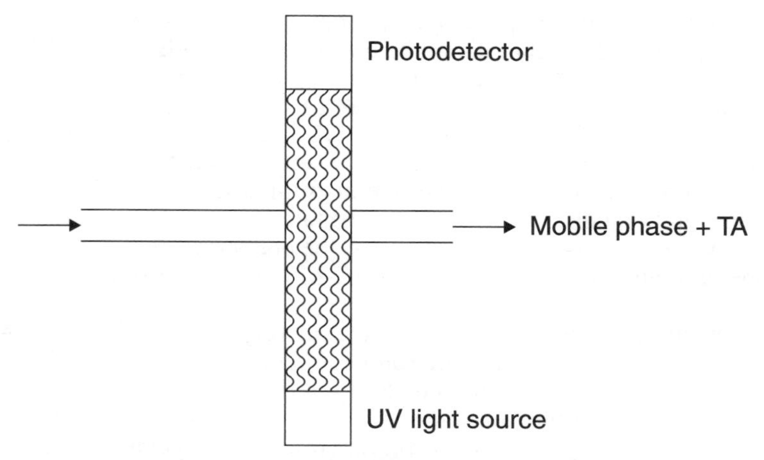

Figure 2-4. Schematic of a UV detector set in a single wavelength mode.

K. Fluorescence

Some molecules are capable of absorbing light at one frequency and becoming excited, then relaxing to a lower excitation state through emission of thermal energy. After a period of time the excited molecule completely relaxes to ground state while simultaneously emitting a photon of light. The polyaromatic hydrocarbons (PAH) of 3-ring or greater size are famous for this behavior. Some atoms, such as mercury also display it. The phenomenon is called fluorescence and lends itself to highly selective and quite sensitive detection schemes. The light of the excitation is always of shorter wavelength (greater energy) than that of the fluorescent emission. The most sensitive configurations use a tuned laser for the excitation source and a broadband photodetector for the measurement component. Fluorescence is akin to absorbance in that a linear relationship governs the calibration. The technique is not free from interferences in that some samples contain co-eluting materials that can quench the fluorescence. The fluorescence detector is non-destructive.

L. IR

The motion of the atoms within a molecule lie in the infrared (IR) frequency range of the electromagnetic spectrum. These motions, predominantly vibrational modes, are capable of absorbing IR light of the same frequency. The intensity of the absorption is not very large, and thus IR monitoring is somewhat limited to determining major components of the injected sample. The sensitivity of the IR analysis under the best of conditions is only in the nanogram on-column range, while most other techniques used in environmental analysis are sensitive at the picogram and lower levels. On the positive side of the ledger, IR spectra are very information rich, and can generate definitive identifications for both calibrated and uncalibrated analytes. Even some of the PAH that are difficult to distinguish by mass spectral analysis are quite clear-cut when IR spectra are obtained. The IR detector is non-destructive.

M. MS

Coupling a mass spectrometer to a gas chromatograph creates the workhorse of the environmental organic analytical instruments. Coupling an MS to an LC creates a beast that has innumerable potential applications that have been essentially ignored by commercial laboratories. It might be that the power and flexibility of the LC-MS serves as the biggest deterrent to its use. Just looking at the coverage that I give in this work to GC and comparing that to my coverage of LC in environmental analysis gives the reader an indication of the predominance of GC in this industry. Since GC totally dominates LC in this industry, then the LC-MS receives almost no attention. This is completely disproportional to the utility of the instruments. GC and GC-MS are mature techniques, essentially tapped-out for environmental. LC and LC-MS are the determinative tools of the next level of analysis; the endocrine disruptor problem can be cited as one particular area where GC and GC-MS have reached their maximum utility.

The mass spectrometer is a destructive detector, very very sensitive, but not very efficient. Of 1000 molecules entering the source, only a very few, generally around one, actually are detected as ions. The sensitivity of the instrument for detecting single ions, however, makes up for the inefficiency of the ionization and sorting process.

The mass spectrometer performs several functions in its operation. The first is to ionize the molecules exiting the column. The most common technique is electron ionization. Electrons from a heated filament collide with the molecules entering the source. The ionized molecules first eject an electron to create a positive ion, then some may decompose creating a daughter ion. It is an axiom of mass spectrometry that only one detectable ion comes from each ionized molecule, be it the molecular ion or a daughter ion. Once the molecular ions and daughter ions are created the fun begins. The ions must be accelerated and focused into a beam that enters the mass analyzer, regardless of whether it is a quadrupole, ion trap, magnetic sector, or time-of-flight analyzer. The analyzer itself pulses and cycles magnetic and electric fields to allow a predetermined series of ions to enter the ion multiplier tube. All of these electrical, magnetic, and time adjustments are what are evaluated in the tune of the mass spectrometer. They are also the inefficient parts of the process. The ion multiplier tube is the efficient part. It's non-specific. Any positive ion entering the tube will be registered as a signal.

Despite all these technical adjustments and requirements to obtain usable data from the mass spectrometer, the quality of the resulting information is well worth the effort. A mass spectrum of a target analyte, and in some cases for a non-calibrated analyte also, can oftimes provide a conclusive identification of the compound, particularly when coupled with elution within a characteristic retention time window. But it is not infallible, and the interpretation of mass spectral results is not for novices.

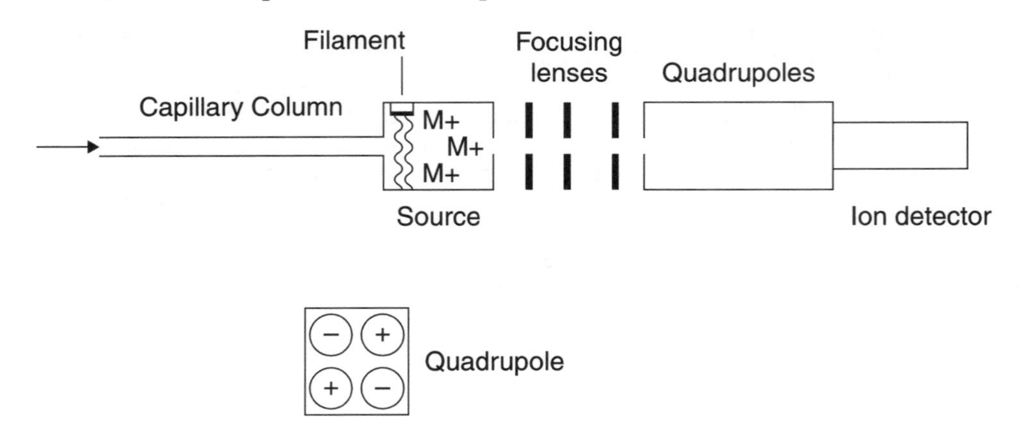

Figure 2-5. Schematic of a quadrupole mass analyzer.

As a rule, if the analyst has an idea of what the unidentified peak is, the mass spectrum of the compound can often provide conclusive identification information. If the analyst has no idea what the peak might be, the mass spectrum can sometimes lead one into completely absurd potential identifications. This text is not intended to be a primer on mass spectral interpretation; there are many competent discussions of mass spectral interpretation that should be consulted[4]. All of these references contain the same limiting assumption that the mass spectrum that is being interpreted is that of one and only one compound. This is not always an applicable assumption in environmental mass spectrometry. Part of the problem is that the mass spectrometer is a general detector. Anything that can come off a column and be introduced into the source will generate signals. If two or more co-eluting compounds are introduced into the source at the same time, the resultant spectrum is going to contain contributions from all the co-eluting compounds and be very confusing to interpret. The mass spectrometer is limited by whatever is presented to it from the separation column. Thus the first step in becoming a good mass spectrometrist is to become a good chromatographer.

There are data manipulation tools that can help the analyst to decide what mass signals are from what eluting component. Background subtraction is a common technique for identifying and removing non-compound mass signals from the spectrum. The analyst obtains a spectrum from a scan within the bounds of the peak profile, generally at the apex of the peak. Next a mass spectrum is obtained from a scan that is off the peak (background). The background spectrum is then subtracted from the spectrum of the peak.

Single ion profiling allows the interpreter to decide what mass signals are coming from what peak. In Figure 2-6 one peak (19.041 minutes) in a cluster of peaks is being examined for potential identification, with the mass spectrum of the peak presented in the lower part of the figure. Without spending a lot of time on the reasons why, the question to be addressed in the interpretation is how much weight needs to be placed on the signals at mass 285 and 287 as compared to the mass signals at 299 and 301. Single ion profiles for these mass signals are presented in Figure 2-7. These profiles demonstrate that the mass signals in question are arising from adjacent peaks rather than being intrinsic to a single peak of interest. Thus they are discounted in the interpretation of the mass spectrum.

[4] Silverstein, R.M., G.C. Bassler, and T.C. Morrill, 1991. *Spectrophotometric Identification of Organic Compounds, Fifth Edition.* John Wiley & Sons, New York, NY.; McLafferty, F.W., and F. Turecek, 1993. *Interpretation of Mass Spectra, Fourth Edition.* University Science Books, Sausalito, CA.; Watson, J.T., 1997. *Introduction to Mass Spectrometry, Third Edition.* Lippincott-Raven, Philadelphia, PA.; Hites, R.A., 1992. *Handbook of Mass Spectra of Environmental Contaminants, Second Edition.* Lewis Publishers, Boca Raton, FL.

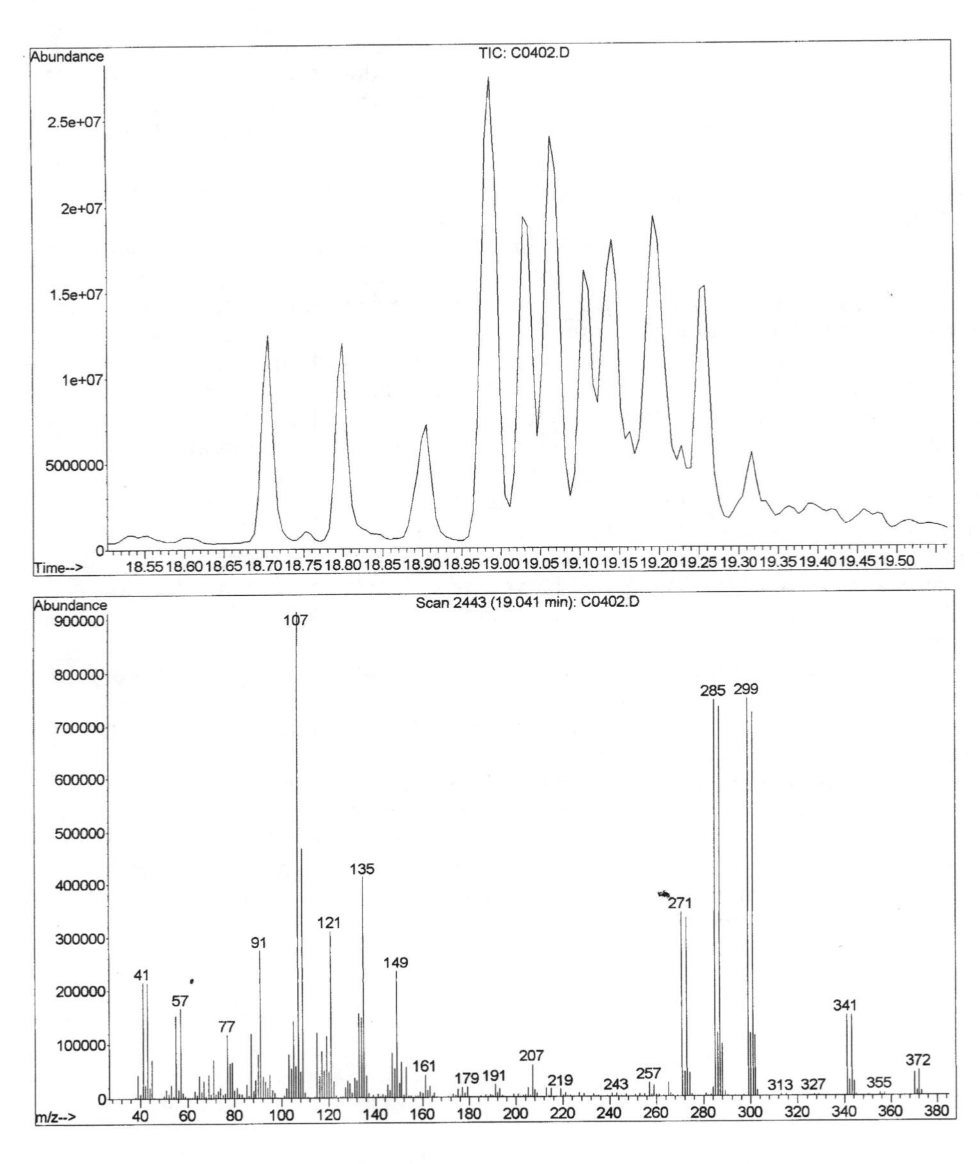

Figure 2-6. Mass spectrum of one peak within a cluster of peaks.

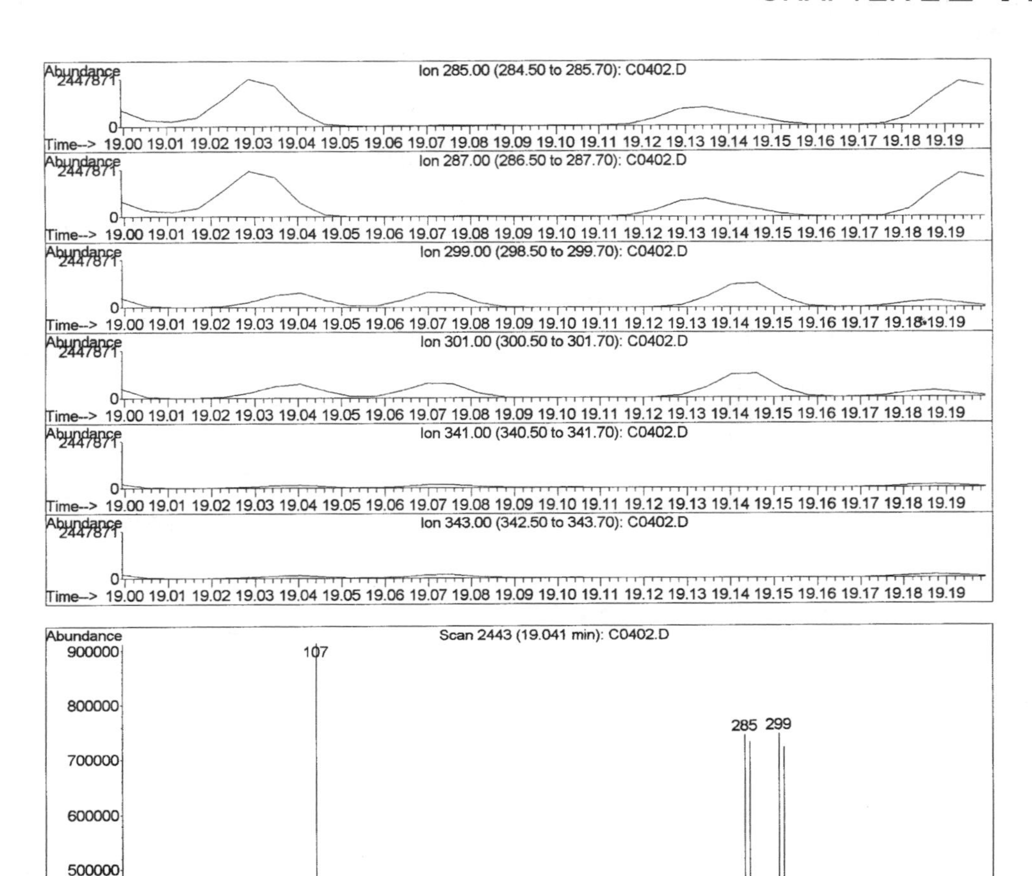

Figure 2-7. Single ion profiles of selected mass signals in the spectrum illustrated in Figure 2-6.

These techniques play an important role in producing a spectrum that can be identified with confidence. The use of detectors for compound identification and quantitation are the subjects of the next two chapters.

CHAPTER 3

Compound Identification

It is a common observation in organic analysis that a correct quantitation can be obtained on a mis-identified analyte. This occurs because many organic compounds exhibit very similar chemical and physical properties, at least as far as the discriminating ability of the detectors employed are concerned. Thus the biggest problem in organic analysis is the conclusive identification of the target analyte.

Correct compound identification has legal ramifications in the environmental industry. For instance, if 1,2-dichlorobenzene is detected in a wastewater effluent, but it is mis-identified and reported as 2-chlorotoluene, the consequence can be fines and possibly court-ordered re-engineering of the manufacturing plant or even shut-down if 2-chlorotoluene is on the discharge permit of the company. This is because discharge permits are specific as to what exact compounds are regulated under the permit. The legal corollary is that if a compound is not listed on the permit, it is not regulated. In many cases when other compounds are found in a particular effluent, and the compounds are not listed on the permit, the permit is interpreted to act as a shield from legal action due to the presence of the non-listed substances.

A. Calibrated Compounds

"Calibrated compounds" refers to those analytes where a standard is on-hand in the laboratory. Further the standard has been used to establish a retention time, a calibration curve, and a detection limit on the instrument that is used for the determination. Identification of the substance boils down to comparison of the responses of the substance in the sample with that of the standard. The emphasis of comparison changes with the different instrumental techniques.

Two-Dimensional Detectors

There are different criteria for target analyte identification, depending on whether a two-dimensional analysis or a three-dimensional analysis has been performed. The only information that is obtained from a two-dimensional analysis is that at a particular retention time, the detector generated a response. None of the two-dimensional detectors in use are specific; and for any given signal the analyst has no way of telling whether evidence of a substance that is highly or only slightly responsive in the detector is being observed. The retention time of a response in any two-dimensional detector system can in no way be considered unique information, it is only one piece information out of a set of information that collectively may allow a conclusive identification.

The example of benzene analysis with a GC-PID was originally examined in Berger.[1] The compounds listed in Table 3-1, were found in the literature to be potential co-elutants with the target analyte benzene. Each of these compounds will also generate a response from the PID, albeit in some cases a very small response. A number of these compounds, particularly ethyl acetate, cyclohexene, 1,2-dimethoxyethane, and 1,3-dioxolane, are in widespread use in industry and can

[1] Berger, W., H. McCarty, and R.-K. Smith, 1996. *Environmental Laboratory Data Evaluation*. Genium Publishing, Schenectady, NY 1-800-243-6486.

reasonably be expected to be present in samples. Without further information then, obtaining a peak within the retention time of benzene from a GC does not guarantee that benzene is present.

Table 3-1. A partial listing of the compounds that can co-elute with benzene during analysis with a GC-PID.

Methyl acrylate	Hexafluorobenzene	Ethyl acetate
Cyclohexene	Cyclohexadiene	Trimethylstilbene
Dichlorocyclohexene	Bromocyclohexene	Methylhexylether
Dimethyl-isobutylamine	2,3-Dichloro-1,4-dioxane	1,3-Dioxolane
Diphosphine	Diisopropylamine	1,2-dichloroethane
Isocyanoethane	Ethylisobutylether	Ethylchloromethylether
Nitroethane	Ethylpropargylether	Bromochloroethene
1,2-Dimethoxyethane	Diisopropylfluorophosphate	Ethylal
Allyl formate	Propyl formate	beta-Furlyl chloride
2-Methyltetrahydrofuran	Hexadiene	Hexatriene
Dichlorohexadiene	Divinylacetylene	3-Methyl-1-hexene
Hexyne	Dichloropropionaldehyde	Pyran
Dimethoxydimethyl silane		

There is a difference legally between reasonable expectation and scientifically rigorous evidence. A frequently heard defense of the GC-PID as a tool for BTEX (benzene, toluene, ethylbenzene, and xylenes) determination is that if the sample was collected from an area where gasoline is suspected to be present, and individual peaks are found in the chromatogram that elute in the benzene, toluene, ethylbenzene, and xylene retention time windows (pattern recognition), the reasonable expectation is that gasoline and BTEX are present in the sample. From an environmentally conservative viewpoint this reasonable expectation of BTEX and gasoline contamination may be sufficient data to go out on the site and begin remediation to remove the purported BTEX. But if the data are intended to be introduced in a court action as evidence of site contamination and the intent of the action is to force an unwilling defendant to pay for the remediation, a reasonable expectation is not equivalent to scientifically rigorous demonstration of the presence of BTEX. Further, just because BTEX may be suspected to be present based on the GC-PID analysis, there is no concrete series of inferential steps that can lead to conclusive determination of gasoline presence. The GC-PID instrument used for the determination is neither calibrated for, nor especially sensitive to gasoline as a target analyte.

Following the directions presented in the EPA GC methods for production of definitive data for compliance monitoring purposes, a laboratory will analyze the sample extract on an analytical column and on a confirmation column. The confirmation column is chosen to exhibit different retention characteristics from those of the analytical column. The retention time order for any given set of target analytes should be different between the analytical column and a suitable confirmation column.

There are so many different columns commercially available now, especially on the LC side of organic analysis, that finding two complementary columns is generally not a problem for at least some of the analytes in the target set. One key point in the choice of suitable columns is to have complete resolution and differentiation of target analytes. It sometimes occurs that a pair of target analytes will co-elute on one column. Although this is not desirable, it happens, and any co-eluting pairs on one column must be completely resolved on the other column for an acceptable confirmation. And this does not mean that compound A can co-elute with compound B on one column, while compounds A and C co-elute on the other column. This is not scientifically rigorous for identification. For some sets of analytes, a suitable confirmation can be achieved only through the use of two different confirmation columns.

The GC instruments normally are configured to have two columns in the same oven with a separate pair of injectors and detectors for each column position. One solution to the dual column confirmation problem is to sequentially inject the sample on first one column then the other. Sample #1 is run on the front column and Sample #2 is run on the rear column at the same time. For the next run, Sample #2 is placed on the front column, and Sample #3 is placed on the rear column, and so forth. This is efficient use of instrument time; however, the analyst is limited to having the same temperature program and run time for both columns.

A less satisfactory solution is to use one injector that is connected with a Y-splitter to two columns, with each column attached to a different detector. A simultaneous analysis and confirmation is recorded for the sample with each injection. Again the choices of temperature programs and run time are limited. Further, the Y-splitter functions best when equal backpressures are present from each column. This means that the columns should have the same length, column diameter, and thickness of liquid layer. The maintenance of the column attachments to the injector can present operational problems and cause results to vary from one day to the next, leading to a loss of analytical consistency.

In either instrument configuration analysts tend to limit their choices of columns to ones that can achieve the analysis in identical times, thereby limiting some of the operational flexibility that can lead to obtaining optimum resolution on each column. This places more demand on the chemical interactions between the analytes and the column liquid layers to achieve the separation.

In LC most of the instrument configurations have only one injector and one detector, which allows a single analysis to take place at a time. To perform confirmation, the analyst must physically remove the analytical column from the instrument and attach the confirmation column in its place. This inherently gives the analyst maximum flexibility in choice of operational conditions, mobile phases, and types of columns. The confirmation column is not limited to merely using a different composition reverse-phase column. Instead, since the analyst is having to re-plumb the instrument anyway, completely different analyte-stationary phase interactions can be used. Examples are using a normal-phase column of silica gel or alumina composition. Separation mechanisms such as gel permeation size sorting or ion-pair chromatography can be employed, sometimes with significant improvements in confirmation ability. A particularly elegant LC confirmation can be achieved when some of the target analytes are composed of optical isomers. Use of a chiral column can generate two identical-size peaks, each with characteristic retention times for an analyte, which improves the quality of the confirmation. Analytes without optical isomers will present only a single peak on the chiral column.

Both columns are completely calibrated for the target analytes with retention time windows established for each of the individual analytes on each column. Elution of a peak at the proper retention time on both columns is considered to be definitive identification for regulatory purposes. It may still not be scientifically rigorous. If the same amount of analyte is determined to be present on both columns, this helps bolster the identification.

This procedure works for many samples if there are few interferences presented by the sample, and a sizable, well resolved peak is present for the analyte. The procedure also works well for determining that the analyte is not present in the sample. If a peak in the appropriate retention time window is exhibited on one column but absent on the other, this constitutes scientifically rigorous evidence that the target analyte is not in the sample extract.

Unfortunately, a common application of the dual column identification technique is on samples that exhibit a considerable amount of background interference, and the maximum sensitivity of the detector is being required. In these samples there is a high probability that there will be fortuitous elution of different components of the background in the appropriate retention time windows on the two columns. This is illustrated schematically in Figure 3-1.

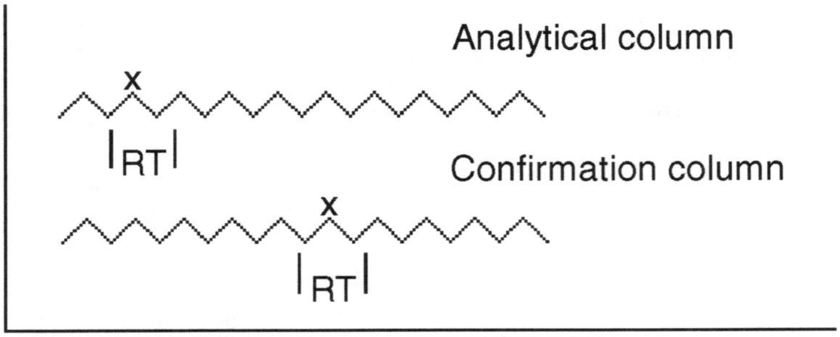

Figure 3-1. Random elution of signals in appropriate retention time windows leading to false positive results.

The amount of the false-positive analyte that is generated in these cases is almost always substantially below the detection limits of alternate techniques (GC-MS) that potentially could be used for identification. The laboratory is normally constrained to report the false-positive as a hit, while common-sense would lead one to view the results as inconclusive and unreliable.

Two-dimensional detectors are more reliable for the determination of multi-peak analytes rather than analytes that elute as a single peak. Examples of analytes of this class include the Arochlors, toxaphene, chlordane, petroleum fuels, and several of the naturally derived pesticides such as the pyrethrins. For these multi-peak analytes, a pattern-recognition process is performed to establish the identity of the analyte. The essential elements of a pattern-recognition identification are the presence of key peaks in appropriate retention time windows, the relative spacing of retention times between the key peaks, the relative abundance of the key peaks to each other, and finally, the width of the pattern as a whole. In some cases sufficient information in these areas can be derived from a single analysis with a two-dimensional detector to conclusively identify the analyte. In other cases, use of a confirmation column or alternate analytical technique may be necessary. The major factors that can lead to identification uncertainty arise from environmental degradation of some, but not all, of the compounds in the multi-component analyte, the presence of co-extracted background interference, and the presence of more that one multi-peak material in the sample.

Environmental degradation is frequently expressed by the term "weathering"; however, this is just one of the processes that can occur. Weathering is most frequently used to describe the loss of the more volatile components from the analyte due to evaporation. It can also encompasses selective destruction/transformation of components of the analyte by outdoor exposure to rain and UV light from the sun. Other causative agents that can result in departures of multi-peak analytes from their normal analytical appearance include bacterial or fungal biodegradation of the analyte and differential adsorption to solids (dirt). Whole body measurements of residues in mammalian and other animal tissues often presents distorted patterns due to metabolism of selected components of the analyte. Detected analytes that have been subjected to degradation can present problems with identification since there may be significant departures from the appearance of the pristine substance, resulting in a reported non-detect when, in fact, the substance may be present. Even when the analyte is correctly identified, an associated problem exists with quantitation as the analytical instrument is obviously not calibrated with the degraded material.

Non-specific co-extracted background interferences (hump-o-grams) can obscure significant parts of the pattern recognition keys. The solution is to apply suitable clean-up procedures to remove the interference. The sulfuric acid clean-up is standard for all PCB samples, but can also be applied to extracts for toxaphene, polychlorinated naphthalenes (PCN, halowaxes), polychlorinated terphenyls (PCT), and polychlorinated alkanes (PCA). PCB, PCN, and PCT can be isolated from hydrocarbon oils and food oils through use of silica gel chromatography. In the case of the hydrocarbon oils, the hydrocarbons elute from the column first. In the case of fatty acids and triglycerides (food oils), the PCB, PCA, and PCT elute first. Chromatography using a styrene-divinyl benzene solid phase (gel-permeation phase) can also be used to isolate PCB, PCA, and PCT from food oils.

The presence of more than one multi-peak analyte can cause some real problems. In the PCB analysis, the presence of multiple Arochlors is essentially a separation problem, and it is most efficiently solved by slowing down the temperature program and increasing the length of the run. However, the Arochlors are not the only multi-peak halogenated analytes that will be encountered. Representatives of other materials include the polychlorinated alkanes[2], the polychlorinated naphthalenes[3], polychlorinated terphenyls, and the polybromonated diphenylethers.[4] Individual compounds within these groups are difficult to selectively remove from the extract and will elute under the same GC conditions as the PCB. Again, the best procedure is to increase the resolution so that it is obvious that the peaks are not falling in the individual PCB retention time windows. Then the laboratory should attempt to identify the materials, generally through GC-MS techniques. The use of a confirmation column can be useful in this situation in verifying that the material is not an Arochlor, but it does not generate much information with regard to what the substance is.

There have been sophisticated attempts to automate the identification process for selected multi-peak analytes.[5] However, an experienced GC analyst is capable of recognizing significantly more data clues and arriving at a correct pattern identification than the average computer integration/identification program, particularly when there are chromatographic interferences present. Visual examination of the chromatogram is the key. Anybody who examines a print-out of retention times and area counts in preference to detailed visual examination of the chromatogram is generally a failure at GC analysis.

The Arochlor standards are quite consistent from one batch to the next and for the most part match up well with the analytes that are found in environmental samples.

---

[2] Tomy, G.T., J.B. Westmore, G.A Stern, D.C.G. Muir, and A.T. Fisk, 1999. Interlaboratory study on quantitative methods of analysis of C10-C13 polychloro-n-alkanes. *Anal Chem.* 71(2):446-451.

[3] Kannan, K., T. Imagawa, A.L. Blankenship, and J.P. Giesy, 1998. Isomer-specific analysis and toxic evaluation of polychlorinated naphthalenes in soil, sediment, and biota collected near the site of a chlor-alkali plant. *Environ. Sci. Technol.* 32(17):2507-2514.

[4] Marsh, G., J. Hu, E. Jakobsson, S. Rahm, and A. Bergman, 1999. Synthesis and characterization of 32 Polybrominated diphenyl ethers. *Environ. Sci. Technol.* 33(17):3033-3037.

[5] Glassmeyer, S.T., K.E. Shanks, and R.A. Hites, 1999. Automated Toxaphene Quantitation by GC/MS. *Anal. Chem.* 71(7):1448-1453

This is not always true for toxaphene, chlordane, and petroleum fuel standards. For these analytes it helps to have a variety of standards from different manufacturers available so that the range of normal variation can be established before attempting to analyze environmental samples.

Three-Dimensional Detectors

The three-dimensional detector, most commonly encountered in the form of the gas chromatograph-mass spectrometer (GC-MS), is not an infallible identification tool; however, it provides significantly more information than a two-dimensional detector. The retention time of the peak is obtained from the detector and is used as one significant item of information for identification of target analytes. The rest of the identification information comes from the spectrum, be it mass (MS), infrared (IR), or ultraviolet-visible (UV-Vis). The majority of this discussion will focus upon the use of MS for identification, but the interpretational techniques are more or less applicable to the other instruments.

Some conventions and terminology in mass spectrometry: The mass spectrum is normally visually presented as a line chart with relative abundance on one axis (vertical) and mass/charge (m/z) on the other (horizontal) as illustrated in Figure 3-2. The largest peak is called the base peak and is assigned the relative abundance of 100. In Figure 3-2 this is at m/z 149. The abundances of all the other signals in the spectrum are reported relative to that of the base peak. There is frequently a signal at an m/z value that corresponds to the molecular mass of the analyte, called the molecular ion; however, it is not commonly the most abundant signal in the spectrum. In Figure 3-2, this peak would be at m/z 390, however it is not visible. Instead the signals at m/z 279 and 280 are the highest seen. All of the m/z signals of lower mass than the molecular ion are called daughter ions.

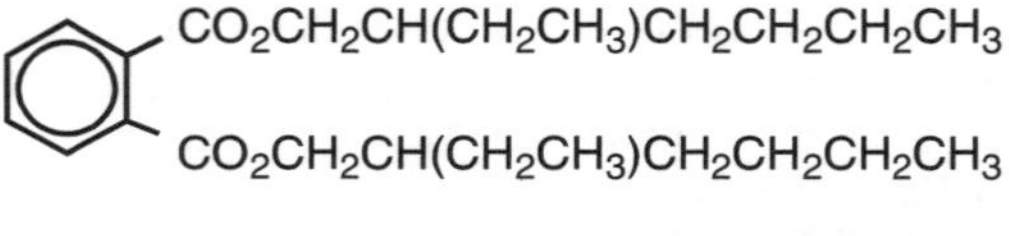

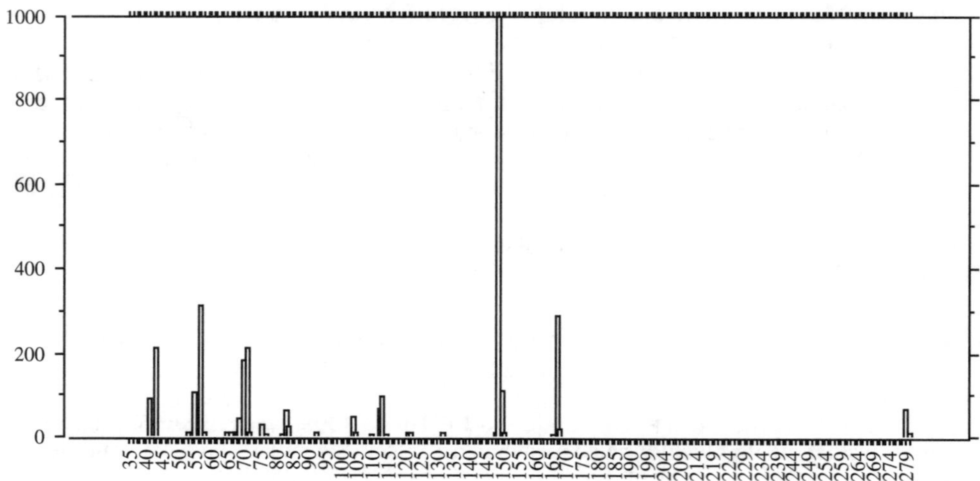

Figure 3-2. Mass spectrum and structure of bis(2-ethylhexyl)phthalate.

The operational software of the mass spectrometer identifies target analytes by examining a retention time window for an indicator m/z signal. The retention time window is normally expressed as a time relative to the retention time of an assigned internal standard. For example, if bis (2-ethylhexyl)phthalate is the target analyte, it is normally assigned to the internal standard chrysene-d_{12}. The absolute retention times of the two compounds are 19.08 and 19.05 minutes, respectively. The relative retention time for bis(2-ethylhexyl)phthalate is 1.0016. The characteristic mass of bis(2-ethylhexyl)phthalate is m/z 149. According to the *National Function Guidelines*[6] if a m/z 149 signal is found within ±0.06 of 1.0016, the analyst can identify the signal as due to bis(2-ethylhexyl)phthalate. Compared to the ability of the instrument this is a huge interval and corresponds to elution of the target compound any time in the range 17.94 to 20.22 minutes. Most GC with a capillary column can maintain an absolute retention time variation of less than 15 seconds and the retention time windows used by the software generally use this ability, although the window sizes can be adjusted by the operator.

Once the software has detected the characteristic m/z signal within the allowed retention time variance, it performs a check of the secondary masses and abundances then rates the confidence of the identification.

Compounds are identified manually by comparison of the mass spectrum of the candidate peak and the mass spectrum of the calibrated compound. This is a two-directional comparison. First the candidate spectrum is evaluated to see if all the significant peaks in the library spectrum are present. It should be kept in mind that mass spectra of members of a class of related compounds may be very similar. For instance the mass spectra of saturated straight-chain hydrocarbons all look for the most part identical. This is illustrated in Figure 3-3, where a pattern of clusters of peaks with central masses at m/z 43, 57, 71, 85, 99, 113, etc. are seen. The single m/z signal that identifies this spectrum as being that of n-hexadecane, as opposed to any of the other saturated straight-chain hydrocabons, is the molecular ion at m/z 226, a signal of relatively minor abundance.

[6] USEPA, October, 1999. *National Functional Guidelines for Organic Data Review*, EPA-540/R-99/008, www.epa.gov/oerrpage/superfund/programs/clp/download/fgorg.pdf

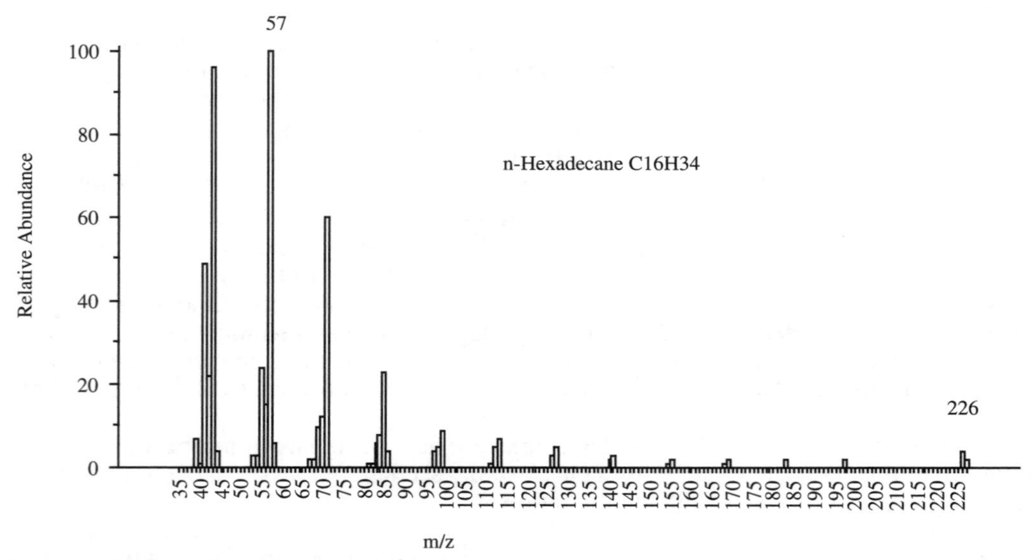

Figure 3-3. Mass spectrum of hexadecane.

After the analyst has determined that all of the significant m/z signals in the library spectrum are present in the candidate spectrum, then the candidate spectrum is examined to see if there are any additional signals present that are not found in the library spectrum. This is actually a very common occurrence. These extra m/z signals may be indicative of a co-eluting compound or they may be germane to the candidate peak. In the latter case they are indications that the candidate is not the target analyte. The technique of single ion chromatograms for evaluating the presence of co-eluting peaks and determining what signals go with which peak was described in the end of Chapter 2. The technique of background subtraction was also mentioned to identify and remove the spectral interferences caused by non-specific signals.

When the candidate peak exhibits m/z signals that are not present in the library spectrum, or vice versa, this is evidence that the compounds may not be identical.

These spectral comparisons are one aspect of organic analysis where computer software is still far behind the abilities of the human interpreter. Most of the programs depend upon matching the m/z that are present in the spectra and the relative abundance of the m/z. These two elements of identification are given about equal weight when the program generates a confidence rating for the match. If one or more masses are missing in the candidate spectrum, the program simply lowers the confidence rating, rather than eliminating the choice as nonsense.

Another shortcoming of the automated identification of target analytes is they do not use all the information that is available in the library spectrum of the target. The operator selects the masses, generally up to 6, that the program will use as identification keys. Since the program is basing its evaluation upon a limited data set, this increases the probability of false positive identifications. The problem is acerbated when the operator picks only the most abundant m/z peaks in the library spectrum, rather than the unique identifying m/z of the compound. When you realize that the program is looking for a particular set of m/z signals within a limited retention time window as identification criteria, this should increase your level of caution in accepting these identifications; however, it frequently only generates a false feeling of confidence by the operator. Within any set of 90^+ target analytes in a calibration, the normal sample will generate 3 to 6 false positive identifications from identification programs. This is a very high rate of false positives, when major management decisions will be based upon the results.

B. Tentatively Identified Compounds (TIC)

The situation is even worse when the retention time window criteria are dropped from the automated spectral evaluation, which occurs when the compound is not a calibrated target analyte. The database of spectra used for comparison is considerably larger that the user prepared library. The NIST library is the largest existing compilation of mass spectra and is the one most commonly imbedded in the GC-MS software. It contains over 75,000 digitized spectra, that were collected from a variety of sources on a variety of instruments under a variety of conditions. In addition to the spectral differences that are to be expected from this situation, there are some mass spectra in the database that are simply mis-identified. These problems, combined with the software's propensity to trivialize the presence or absence of m/z signals, means that identifications of compounds based solely upon the results of a computer library search using a matching algorithm can be wildly unreliable.

Which is not to say that the software-based identifications are wrong 100% of the time. As the saying goes, "Even the habitually incompetent can, by random choice, get something right once in a while." However, it takes a trained and skilled mass spectral interpreter to sift through the computer search results and come up with a pared-down list of what is probable and eliminate what is nonsense.

There are certain groups of compounds that are, on occasion, reported as tentatively identified compounds but should never be considered as potentially present in samples. These groups include:

1. Internal standards and surrogates from the analysis
2. Target analytes of the analysis
3. Target analytes of other analytical methods that were performed on the sample
4. Degradation products from the GC column
5. Laboratory contaminants.

The internal standards and surrogates were intentionally added to the sample by the laboratory as part of the analytical procedure, and are obviously not characteristic of the field source. Target analytes should be handled as reported results from calibrated compounds, rather than as tentatively identified compounds. Item 3 can be qualified to say that if a target analyte of another method is detected, it is probably beneficial to the client to go ahead and verify that the compound is present by performing the additional analysis.[7] For example if a pesticide like diazinon, which is a target analyte of the N-P pesticide Method 8141, is detected in a TIC scan, the analyte can be verified by re-analysis of the extract on a Method 8141 instrument.

The liquid layer on the inside of the capillary columns is a silicone, substituted with a variety of organic moieties. These silicones are sensitive to acids, bases, water, and oxygen, particularly at the elevated temperatures that most GC analyses reach. The breakdown products of the silicones are smaller silicones, typically cyclic materials where there are alternating silicon and oxygen atoms making up the ring. An example of these column-bleed materials is hexamethylcyclotrisiloxane (CAS 541-05-9). Hexamethylcyclotrisiloxane is the smallest of a series of cyclosiloxanes that are created from liquid layer degradation, and is the one most frequently encountered. This

[7] For laboratories that are totally focused only on the bottom-line profit, the client would need to be contacted prior to the re-analysis to insure that they will pay for the re-analysis.

silicone has a low boiling point (134 °C) and a relatively high vapor pressure. It normally forms continuously in the column and contributes to the undifferentiated background. In certain situations, particularly in columns that are rapidly degrading due to introduction of acids and water, the formation can occur at low temperatures. Then, when the column is heated, the breakdown products can elute as distinct peaks.

The GC columns used in environmental analysis by MS will almost always have liquid phases that are 95-100% methyl-substituted. The methyl-substituted cyclosilicones are characterized by m/z signals at 207, 281, and 355, arising from loss of a CH_3- from the molecular ion. Figure 3-4 is a mass spectrum of hexamethylcyclotrisiloxane. Close examination of almost any mass spectrum generated from a GC-MS will reveal m/z signals at these masses, particularly m/z 207.

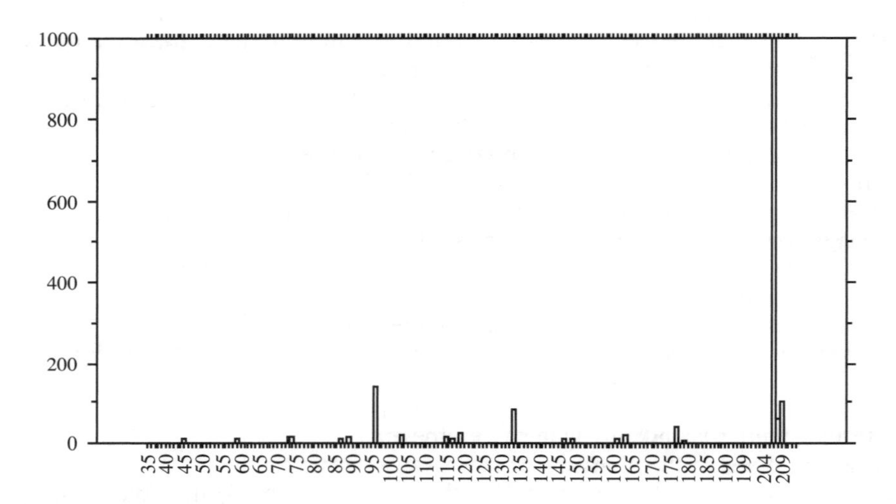

Figure 3-4. Mass spectrum of hexamethylcyclotrisiloxane.

Silicones generally have very distinct mass spectra, and the library matching software will frequently correctly identify the peak. But, in the vast majority of situations, they are not in the sample. These materials are too sensitive to exist for any length of time in the environment. They are merely pauses on the way to becoming sand (silicon dioxide). Even in samples collected from a silicone manufacturing facility, there is a very low probability that cyclosiloxanes will be present.

There once was a man that who came to my laboratory. He was all excited because another lab had analyzed a sample of gas he had collected from a vent pipe on a landfill and reported hexamethylcyclotrisiloxane. He had looked in the Aldrich catalog and found that hexamethylcyclotrisiloxane was selling for about a dollar a gram. He was in the process of setting up a condenser at the landfill to collect the hexamethylcyclotrisiloxane, and wanted my lab to purify the material so he could sell it. I tried to explain to him that it was a false positive and what was being reported was column bleed, an artifact of the analysis. He countered by showing me the chromatogram and the mass spectrum. The chromatogram exhibited a major rise in the baseline throughout the run, good evidence of column degradation. But the man could not be swayed. He had his gold mine and was adamant in his needs. I finally told him that unfortunately my laboratory did not have the facilities to do the purification and suggested he try another laboratory that might be able to help him.

Laboratory contamination will be discussed in detail in Chapter 7; however, it is important when evaluating TIC. The primary tool for evaluating laboratory contamination is the blank. The laboratory should pay considerable attention to the results of the blank with regards to the calibrated target analytes, especially the phthalates and the common laboratory solvents such as acetone and methylene chloride. However, anything that is not calibrated is generally ignored, and further, will not show up on the quantitation report from the blank. But these non-calibrated substances present considerable problems when TIC are requested.

There are many examples of these frequently found laboratory contaminants. One is butylated hydroxy toluene (BHT, 2,6-bis[1,1-dimethyethyl]-4-methylphenol, CAS 128-37-0), the primary member of a family of compounds used as anti-oxidants in foods and plastics to prevent discoloration and degradation due to oxidation. Other food-derived contaminants can include fatty acids and esters, being introduced into the sample by technicians who do not wash their hands after eating.[8]

Acetone and isopropanol used in field decontamination procedures and laboratory glassware washing can contribute artifacts. Acetone is the oxidation product of isopropanol, and all bottles of isopropanol contain some acetone, unless an anti-oxidant has been added. Acetone has many desirable properties, such as being a powerful solvent for organic substances as well as being miscible with water; however, is not chemically inert. In the presence of traces of acid or base, acetone will react with itself to form condensation products that remain on the glassware or equipment after the acetone has evaporated. The major contaminants that come from acetone are diacetone alcohol (4-hydroxy-4-methyl-2-pentanone, CAS 123-42-2) and mesityl oxide (4-methyl-3-pentene-2-one, CAS 141-79-7), although other 6-carbon oxygenated compounds can be present as well as higher weight condensation products.

There are other non-calibrated materials that will show up in the laboratory as contaminants on occasion. The point is that these substances are present in the sample extract as laboratory artifacts, rather than being germane to the field source of the sample.

The proper technique is to run a TIC scan on the blank in addition to that on the sample, then perform a one-to-one comparison of the results to eliminate those peaks that are present in the blank. The laboratory should take one of two tacks in the event of laboratory artifact TIC. The first is to delete mention of the TIC from the report. The other is to include them in the report but flag the TIC with a "B" to indicate that it is present in the blank. The first solution is probably less confusing to clients; however, the second is the more complete and scientifically defensible, particularly when instrument raw data are being included for evaluation with the report.

[8] You can always recognize the true "lab rats" in the analytical chemistry laboratory: they wash their hands before going to the bathroom, rather than after.

CHAPTER 4

Quantitation

After the analyte has been conclusively identified, and "conclusively identified" must be stressed, the next step is to determine how much of it is present in the sample. In general terms this is nothing more than a comparison of the response of the compound in the sample extract to the response of standards of known concentration. As with most things, the difficulty lies in the details of the implementation. Details that can have significant influence upon the results include, how the response of the analyte is defined and determined, the technique of calibration, the boundaries of the calibration, and the sensitivity of the instrument used in the analysis.

A. Definition and Determination of Analyte Response

All instruments will generate a background signal. On top of this innate background will be added a background signal due to the sample. Sometimes the sample background is insignificant; however, in most cases it is substantial. The signals due to the analyte are superimposed upon these combined background signals. This is illustrated in Figure 4-1.

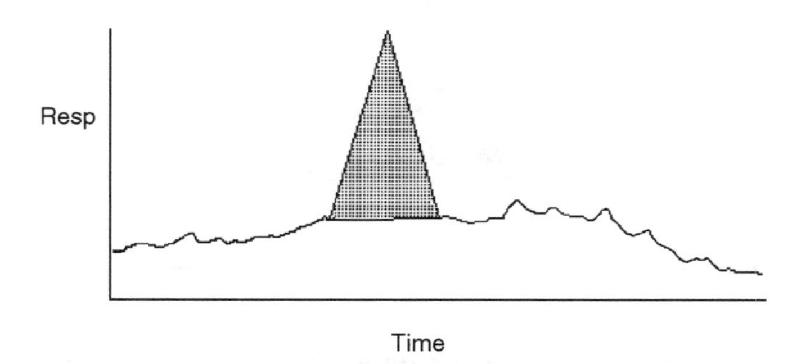

Figure 4-1. Correctly assigned area due to an analyte superimposed on sample background.

The correct procedure to use in assigning peak area is to determine the beginning and end of the peak, then draw a connecting line between these two points. The line constitutes an approximation of the average behavior of the background and should be contiguous with the average background prior to and after the peak. The assigned beginning and end of the peak are where the signal trace leaves and returns to the background.

There are a number of area assignment procedures that are incorrect. The first is to include the background with the peak area, as illustrated in Figure 4-2. This is inappropriate and serves to falsely inflate the size of the peak. It occurs by drawing the background from points other than the beginning and end of the peak. A grossly inappropriate practice is when the beginning and end of the chromatographic run are chosen for the background establishment. An even worse situation is when the resting signal from the instrument is chosen for background.

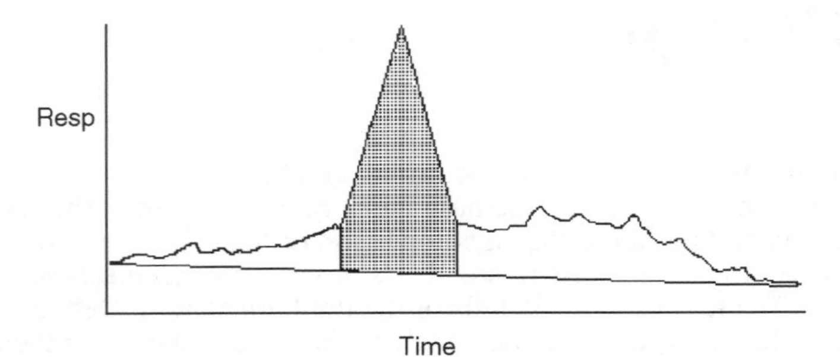

Figure 4-2. Improper procedure I - Inclusion of background in peak area determination.

The second incorrect practice occurs when the beginning of the peak is assigned after the signal trace has left the background, or the end of the peak is designated before the return to sample background. An example of this practice is illustrated in Figure 4-3. A variation of the practice that accomplishes the same result is illustrated in Figure 4-4. Either practice results in too little area being assigned to the peak.

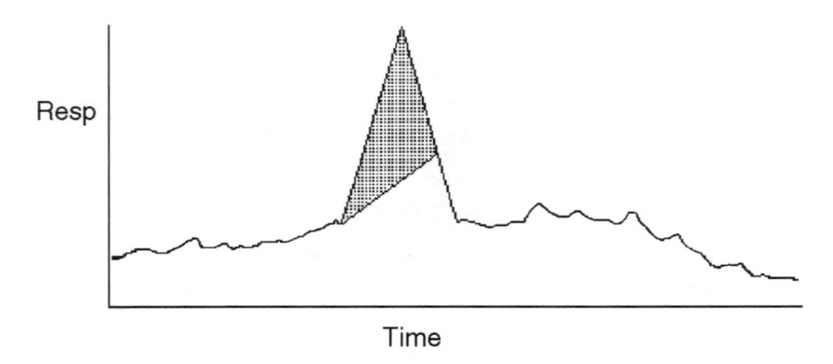

Figure 4-3. Improper procedure II - Termination of the peak before the return to baseline.

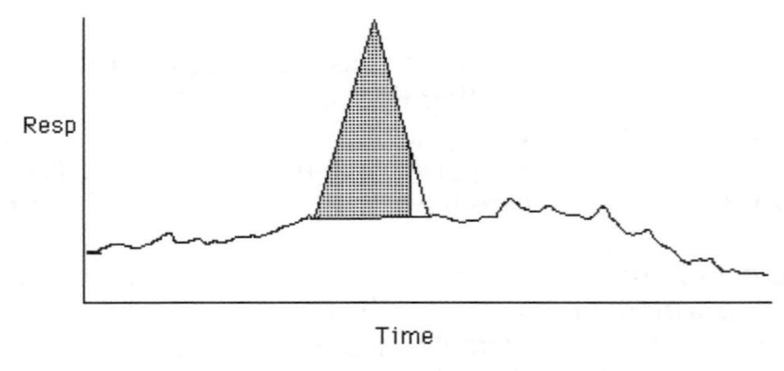

Figure 4-4. Variation on improper procedure II.

The third incorrect practice is performed by extending either or both the beginning and end of the peak to positions beyond the return to baseline. This results in the peak having too much area, as illustrated in Figure 4-5. Often this practice is rationalized by using the boundaries of the retention time window as the peak beginning and end and then saying that any area within the window should be assigned to the peak.

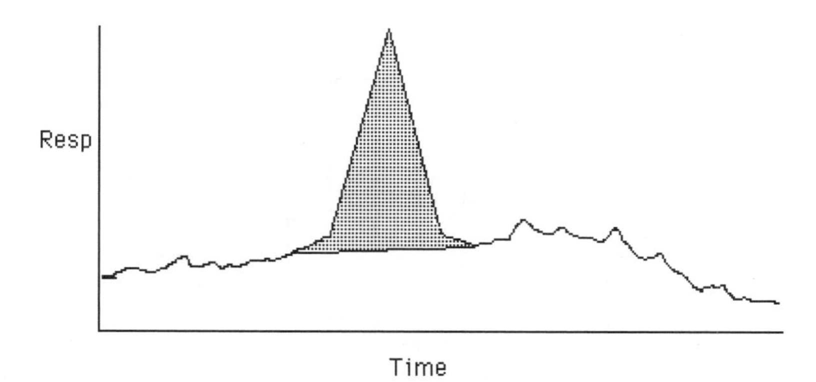

Figure 4-5. Improper procedure III - Extending assignment of beginning and end of peak beyond the return to baseline.

These three practices are what are collectively termed "peak shaving." They are most commonly used to enhance the area of the peak to meet an established quality control recovery limit and thereby avoid re-analysis or re-extraction of the sample or repeating a calibration. The same three improper procedures are encountered regardless of whether the calibration is based on peak area or peak height. In all instances these practices are unconscionable and frequently constitute definitive evidence of laboratory fraud.

Data processing software uses digital representations of the chromatograms. Digital data are one-dimensional data. Each point consists of a value representing the average strength of signal in the time interval. The position of the datum point in the stream of data received by the computer serves to assign a time value to the point, based on the data-sampling frequency. The chromatogram is reconstructed by sequentially plotting the data values. Figure 4-6 illustrates a peak represented by digital data.

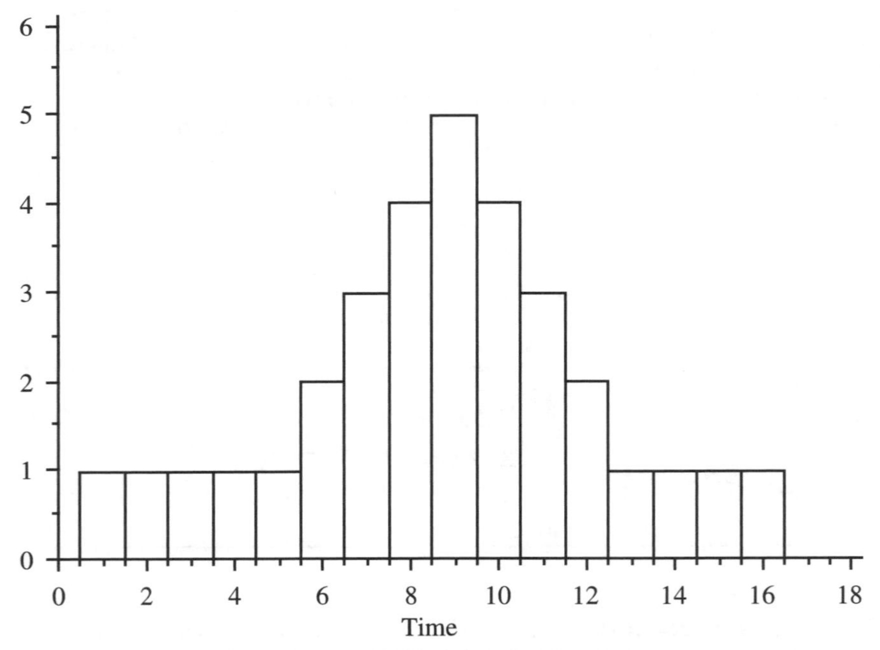

Figure 4-6. Digital representation of chromatographic data.

Most software used for data analysis have automated peak identification and integration capabilities. There are a variety of settings in the software that control how a peak is defined and what area is included in the integration. Peak initiation is assigned by a pre-determined number of consecutive data points that are rising in value. There is normally a minimum rise in value that must be achieved between any two consecutive points before an increase in value is recognized. The end of the peak is assigned by a pre-determined number of consecutive data points that are of relatively constant value. The number of required consecutive points being evaluated will vary depending on the data sampling frequency and the allowed minimum peak width.

For example, let's assume the required number of consecutive points in the buffer is 5 and the minimum difference is 4. Values being collected in the buffer could be 1, 3, 2, 5, 1, 3, 4, 9, 50, 150, 300, 650, 1400. The time slices associated with the values 4, 9, 50, 150, and 300 fit the criteria for 5 consecutive points that have increasing values with differences greater than 4, so the time slice associated with the value 4 is assigned as the peak beginning. At the other end of the peak signals collected could be 1400, 650, 300, 150, 100, 50, 10, 5, 4, 3, 1, 4, 2, 3, 1. The time slice associated with the value 5 begins the series of consecutive points that meet the criteria for baseline and would be designated as the end of the peak. Then the software constructs a baseline between the assigned beginning and end of the peak as in Figure 4-1.

Baseline-to-baseline peak assignment works well when only a few components of interest are in the sample and the chromatogram exhibits few interferences and a relatively stable baseline. This is not the situation for most extracts of environmental samples, where co-extracted interferences and numerous analytes are the norm. An integration technique called valley-to-valley is frequently more useful in these situations. Valley-to-valley integration includes the recognition capability of baseline-to-baseline, but it also allows recognition of a valley between two closely eluting peaks.

Returning to the above example's criteria, suppose the received data are 1400, 650, 300, 150, 100, 75, 100, 150, 300, 650, 1400. The time slice associated with the value of 75 would be designated as a valley point between two peaks. This means that it is simultaneously the end of one peak and the beginning of the next peak.

There are problems in quantitation associated with this type of peak recognition. The software will normally establish baselines from the peak beginning to the valley point and then down to the second peak's return to baseline, as illustrated in Figure 4-7. This results in areas that are too low for each of the peaks. A more appropriate baseline construction in this situation is illustrated in Figure 4-8.

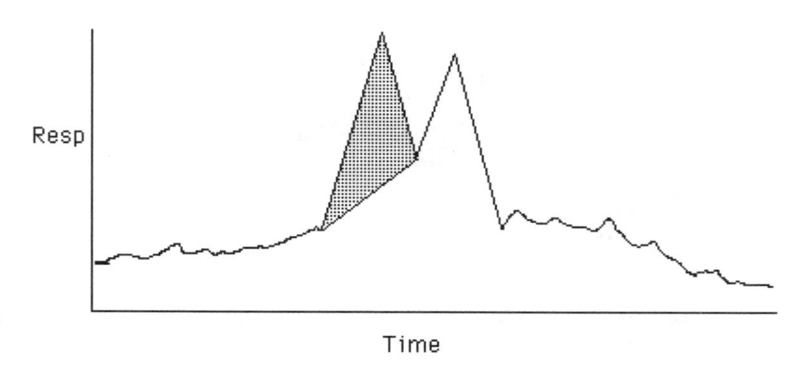

Figure 4-7. Inappropriate baseline construction when valley points are present.

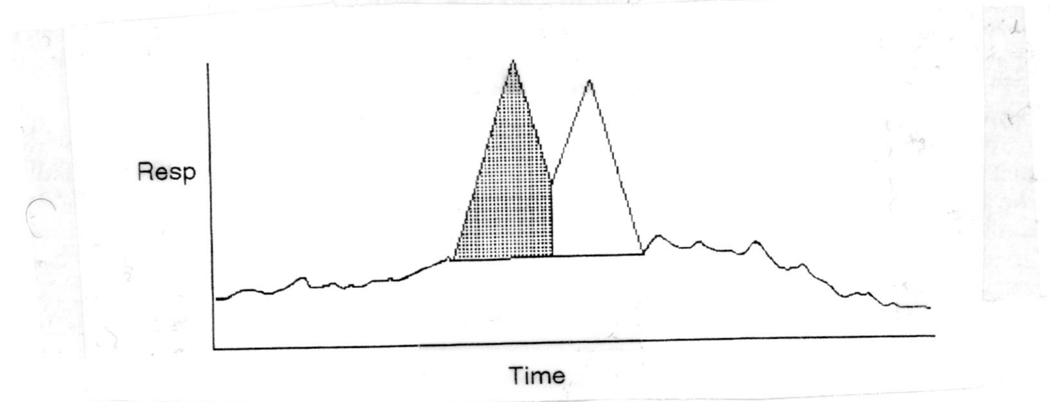

Figure 4-8. Correct baseline construction when valley points are present between peaks.

The baseline construction illustrated in Figure 4-8 is also appropriate when analytes with multiple peaks are being quantitated. For example, if the analyte is diesel fuel, the most reliable area determination is all the area above the baseline from the beginning of the pattern at decane (C_{10}) to the end of the envelope at tetracosane (C_{24}). The retention time markers of nonane (C_9) and pentacosane (C_{25}) are used to bracket the analyte pattern, as illustrated in Figure 4-9. The baseline should be drawn between the end of the nonane peak and the beginning of the pentacosane peak.

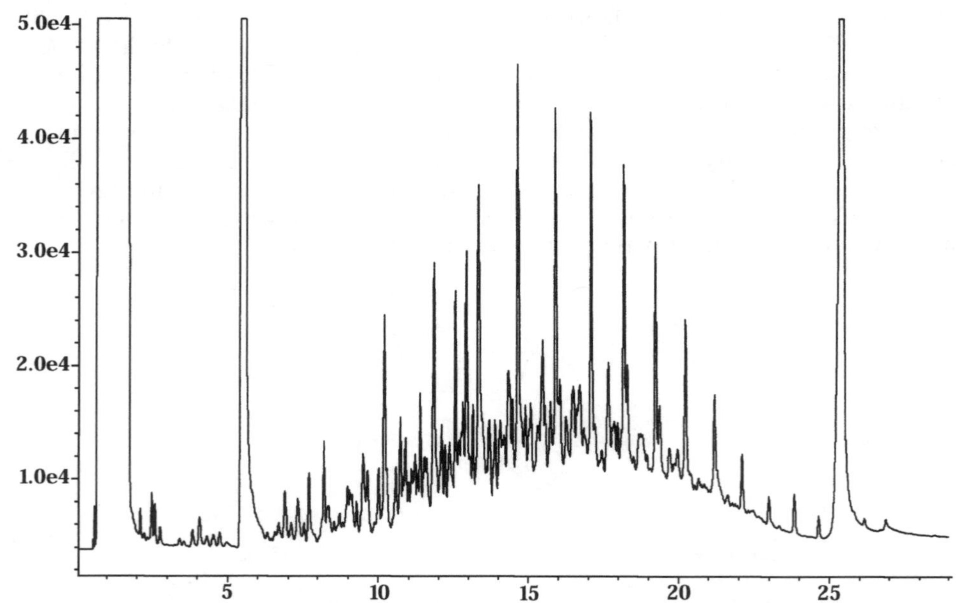

Figure 4-9. Diesel fuel bracketed by the retention time markers nonane and pentacosane (big signals at 5 and 25 minutes).

Manual assignment of peak beginnings/ends and baseline construction is necessary in some instances to correct the results of automated assignments. However, there is a very distinct difference between corrections to obtain appropriate quantitation and peak shaving. It is necessary to document all changes that are made by the analyst from the initial results of the automated integration so that data reviewers can distinguish between these different intents. This documentation would normally consist of a hardcopy print-out of the original results attached to the hardcopy of the final results that include all the manual adjustments. The integration software also keeps track of all the automated peak assignments and integrations, and manual changes that were made to the original results in a file that is referred to as an audit trail. An example of the contents of an audit trail file are presented in Table 4-1.

Table 4-1. Example of an audit trail file.

Entry	Text
(Header)	D:\HPCHEM\1\DATA\020399\A5057.D\Audit.txt Created Wed Feb 03 18:08:36 1999
1	Modified : Wed Feb 03 18:08:36 1999 Event : Quantitation Message : Calculation using continuing calibration QuantFile: 8270625.RES User : Sequence: Why : Sequence Severity : 0
2	Modified : Wed Feb 03 18:08:43 1999 Event : Report Generation Message : Generated Report using option: S QuantFile: 8270625.RES User : Sequence: Sequence: Why : Sequence Severity : 0

Table 4-1. Example of an audit trail file, *continued*

Entry	Text
3	Modified : Wed Feb 03 18:08:45 1999 Event : ISTD Area Message : 262850 1,4-Dichlorobenzene-d4 QuantFile: 8270625.RES User : Sequence: Sequence: Sequence: Why : Sequence Severity : 1
4	Modified : Wed Feb 03 18:08:45 1999 Event : ISTD Area Message : 978069 Naphthalene-d8 QuantFile: 8270625.RES User : Sequence: Sequence: Sequence: Sequence: Why : Sequence Severity : 1
5	Modified : Wed Feb 03 18:08:46 1999 Event : ISTD Area Message : 476296 Acenaphthene-d10 QuantFile: 8270625.RES User : Sequence: Sequence: Sequence: Sequence: Sequence: Why : Sequence Severity : 1
6	Modified : Wed Feb 03 18:08:46 1999 Event : ISTD Area Message : 820001 Phenanthrene-d10 QuantFile: 8270625.RES User : Sequence: Sequence: Sequence: Sequence: Sequence: Sequence: Why : Sequence Severity : 1
7	Modified : Wed Feb 03 18:08:47 1999 Event : ISTD Area Message : 722112 Chrysene-d12 QuantFile: 8270625.RES User : Sequence: Sequence: Sequence: Sequence: Sequence: Sequence: Sequence: Why : Sequence Severity : 1
8	Modified : Wed Feb 03 18:08:47 1999 Event : ISTD Area Message : 567518 Perylene-d12 QuantFile: 8270625.RES User : Sequence: Sequence: Sequence: Sequence: Sequence: Sequence: Sequence: Why : Sequence Severity : 1
9	Modified : Wed Feb 03 18:08:47 1999 Event : Surrogate Recovery Message : 2-Fluorophenol QuantFile: 8270625.RES User : Sequence: Sequence: Sequence: Sequence: Sequence: Sequence: Sequence: Why : Sequence Severity : 1

Table 4-1. Example of an audit trail file, *continued*

Entry	Text
10	Modified : Wed Feb 03 18:08:47 1999 Event : Surrogate Recovery Message : 0.45 Phenol-d5 QuantFile: 8270625.RES User : Sequence: Sequence: Sequence: Sequence: Sequence: Sequence: Sequence: Why : Sequence Severity : 1
11	Modified : Wed Feb 03 18:08:48 1999 Event : Surrogate Recovery Message : 0.00 Nitrobenzene-d5 QuantFile: 8270625.RES User : Sequence: Sequence: Sequence: Sequence: Sequence: Sequence: Sequence: Why : Sequence Severity : 1
12	Modified : Wed Feb 03 18:08:48 1999 Event : Surrogate Recovery Message : 0.00 2-Fluorobiphenyl QuantFile: 8270625.RES User : Sequence: Sequence: Sequence: Sequence: Sequence: Sequence: Sequence: Why : Sequence Severity : 1
13	Modified : Wed Feb 03 18:08:49 1999 Event : Surrogate Recovery Message : 2,4,6-Tribromophenol QuantFile: 8270625.RES User : Sequence: Sequence: Sequence: Sequence: Sequence: Sequence: Sequence: Why : Sequence Severity : 1
14	Modified : Wed Feb 03 18:08:49 1999 Event : Surrogate Recovery Message : 0.00 Terphenyl-d14 QuantFile: 8270625.RES User : Sequence: Sequence: Sequence: Sequence: Sequence: Sequence: Sequence: Why : Sequence Severity : 1
15	Modified : Thu Feb 04 13:36:22 1999 Event : Quantitation Message : Calculation using initial calibration QuantFile: ORANGE.RES Severity : 0
16	Modified : Thu Feb 04 13:36:26 1999 Event : Report Generation Message : Generated Report using option: S QuantFile: ORANGE.RES Severity : 0
17	Modified : Thu Feb 04 13:36:26 1999 Event : ISTD Area Message : 978069 Naphthalene-d8 QuantFile: ORANGE.RES Severity : 1

Table 4-1. Example of an audit trail file, *continued*

Entry	Text
18	Modified : Thu Feb 04 13:36:27 1999 Event : ISTD Area Message : 476296 Acenaphthene-d10 QuantFile: ORANGE.RES Severity : 1
19	Modified : Thu Feb 04 13:36:27 1999 Event : ISTD Area Message : 820001 Phenanthrene-d10 QuantFile: ORANGE.RES Severity : 1
20	Modified : Thu Feb 04 13:36:27 1999 Event : ISTD Area Message : 726591 Chrysene-d12 QuantFile: ORANGE.RES Severity : 1
21	Modified : Thu Feb 04 13:36:27 1999 Event : ISTD Area Message : 567518 Perylene-d12 QuantFile: ORANGE.RES Severity : 1
22	Modified : Thu Feb 04 13:36:27 1999 Event : Surrogate Recovery Message : -0.02 Nitrobenzene-d5 QuantFile: ORANGE.RES Severity : 1
23	Modified : Thu Feb 04 13:36:28 1999 Event : Surrogate Recovery Message : -0.02 2-Fluorobiphenyl QuantFile: ORANGE.RES Severity : 1
24	Modified : Thu Feb 04 13:36:28 1999 Event : Surrogate Recovery Message : -0.03 Terphenyl-d14 QuantFile: ORANGE.RES Severity : 1
25	Modified : Thu Feb 04 13:44:53 1999 Event : Manual Integration Message : Changed peak amount for gamma-BHC from 14.4541ng/uL to 9.56642ng/uL QuantFile: ORANGE.RES Severity : 1
26	Modified : Thu Feb 04 13:45:02 1999 Event : Manual Integration Message : Changed peak amount for Parathion from 13.5153ng/uL to 10.2011ng/uL QuantFile: ORANGE.RES Severity : 1
27	Modified : Thu Feb 04 13:45:06 1999 Event : Manual Integration Message : Changed peak amount for Aldrin from 15.642ng/uL to 8.80706ng/uL QuantFile: ORANGE.RES Severity : 1

Table 4-1. Example of an audit trail file, *continued*

Entry	Text
28	Modified : Thu Feb 04 13:45:09 1999 Event : Manual Integration Message : Changed peak amount for Aldrin from 8.80706ng/uL to 10.0781ng/uL QuantFile: ORANGE.RES Severity : 1
29	Modified : Thu Feb 04 13:45:18 1999 Event : Manual Integration Message : Changed peak amount for Terphenyl-d14 from 14.1321ng/uL to 10.5818ng/uL QuantFile: ORANGE.RES Severity : 1
30	Modified : Thu Feb 04 13:45:27 1999 Event : Manual Integration Message : Changed peak amount for Dieldrin from 14.3093ng/uL to 11.2269ng/uL QuantFile: ORANGE.RES Severity : 1
31	Modified : Thu Feb 04 13:45:36 1999 Event : Manual Integration Message : Changed peak amount for 4,4'-DDD from 15.1265ng/uL to 9.07945ng/uL QuantFile: ORANGE.RES Severity : 1
32	Modified : Thu Feb 04 13:45:43 1999 Event : Manual Integration Message : Changed peak amount for Endosulfan II from 14.0627ng/uL to 8.1035ng/uL QuantFile: ORANGE.RES Severity : 1
33	Modified : Thu Feb 04 13:45:45 1999 Event : Manual Integration Message : Changed peak amount for Endosulfan II from 8.1035ng/uL to 9.56213ng/uL QuantFile: ORANGE.RES Severity : 1
34	Modified : Thu Feb 04 13:45:52 1999 Event : Manual Integration Message : Changed peak amount for Endrin Aldehyde from 13.5541ng/uL to 9.91463ng/uL QuantFile: ORANGE.RES Severity : 1
35	Modified : Thu Feb 04 13:45:59 1999 Event : Manual Integration Message : Changed peak amount for Kepone from 19.1081ng/uL to 12.9935ng/uL QuantFile: ORANGE.RES Severity : 1
36	Modified : Thu Feb 04 13:46:06 1999 Event : State Change Message : From Not Reviewed to QT Reviewed QuantFile: ORANGE.RES Severity : 1

Table 4-1. Example of an audit trail file, *continued*

Entry	Text
37	Modified : Thu Feb 04 13:49:00 1999 Event : Manual Integration Message : Changed peak amount for Kepone from 12.9935ng/uL to 5.24502ng/uL QuantFile: ORANGE.RES Severity : 1
38	Modified : Thu Feb 04 13:49:01 1999 Event : Manual Integration Message : Changed peak amount for Kepone from 5.24502ng/uL to 8.06603ng/uL QuantFile: ORANGE.RES Severity : 1
39	Modified : Thu Feb 04 13:49:03 1999 Event : Manual Integration Message : Changed peak amount for Kepone from 8.06603ng/uL to 9.6428ng/uL QuantFile: ORANGE.RES Severity : 1

The audit trail is a text file that sequentially lists all the integration changes that have been made to a file, beginning with the original automated software integration and report generation (entries 1 through 24 in Table 4-1). Evidence that inappropriate manipulations that have been performed on peak integrations to obtain "within limits" results are present in the audit trail. These key entries are normally labeled "manual integration." The evidence consists of a series of entries for a single compound where there is a sequential increase or decrease in the integrated amount of the peak as the analyst tries to meet some acceptance threshold. As discussed above, there are frequent instances where the beginning of a peak, end of a peak, or baseline adjustments need to be made to the original software assignments. However, these are made only once, and then the software recalculates the result, giving only one manual integration entry in the audit trail. A series of audit trail entries, such as entries 35, 37. 38, and 39 for kepone in Table 4-1, indicates that the instrument operator is seeking to adjust the peak integration incrementally to some pre-determined level - such as the acceptance limit for a quality control measure, or as is the case in this example, a ± 20% D on a continuing calibration standard.

The audit file is located with the raw data file in the sample data directory. The easiest way to find and access the file is to first obtain the raw data file name, then, go to the computer desktop and search down through the directory and file structure until the raw data file is found. The associated audit trail is located in the directory of the raw data. The audit trail can be deleted from the computer through the normal process of file deletion; however, the absence of the audit trail is in itself an indication that unwarranted manipulations may have been performed on the sample data.

B. Technique of Calibration

There is no such thing as a universal calibration in organic instrumental analysis. Each instrument must be individually calibrated with the target analytes. There are no known examples where a calibration is generated on one instrument and is then transferred to another instrument, and the second instrument subsequently produces reliable results.

There are two types of calibration used in organic instrumental analysis. The two types of calibration are called external standard and internal standard. In either case known concentrations of target analyte are analyzed and the associated instrument responses recorded. A mathematical relationship between the concentration and response is generated. The mathematical relationship is called a calibration curve, regardless of whether it is linear or actually exhibits a curve.

The relationship may take a number of different mathematical forms; however, one requirement is that there must be a regular change in response from the instrument with increasing concentration of analyte. At no point in a valid calibration can two different concentrations of analyte generate the same response. Suitable calibration models that are frequently used include:

- Linear through the origin
- Linear not through the origin
- Second degree curves
- Third degree curves.

The linear models are based upon the standard two-dimensional equation for a straight line,

$$y = mx + b$$

where y is the amount of analyte, x is the instrument response, m is a constant equivalent to the slope of the line, and b is the intercept. If the line goes through the origin, then the value for constant b is zero. If b is any value other than zero, then the line does not go through the origin, and the associated calibration is linear not through the origin. Any two points can be connected by a straight line. Thus to demonstrate that a linear equation is a suitable model for a particular calibration, at least three different concentrations of analyte are needed. The zero point (blank) is not a reliable point for calibration purposes.

The second degree curves are represented by equations containing terms where the response is raised to the power of two. A general equation for such a curve is,

$$y = ax^2 + bx + c$$

where y is again the concentration of the analyte; x is the instrument response; and a, b, and c are constants. The constant c is the intercept. If c is equal to zero, then the curve passes through the origin. Values of c other than zero mean the curve does not pass through the origin. Any three points can be connected using a second-degree equation. To demonstrate that a second-degree equation is suitable for a particular calibration, at least four different concentrations must be analyzed. Even when four points are included in the calibration the analyst must visually examine the graph of the curve to verify that there are unique values for each concentration level. An example of an unacceptable curve is illustrated in Figure 4-10. In this example, the correlation coefficient is greater than .999, indicating that there is a very high level of accurate mathematical description, however a simple visual examination shows the relationship to be worthless for most purposes.

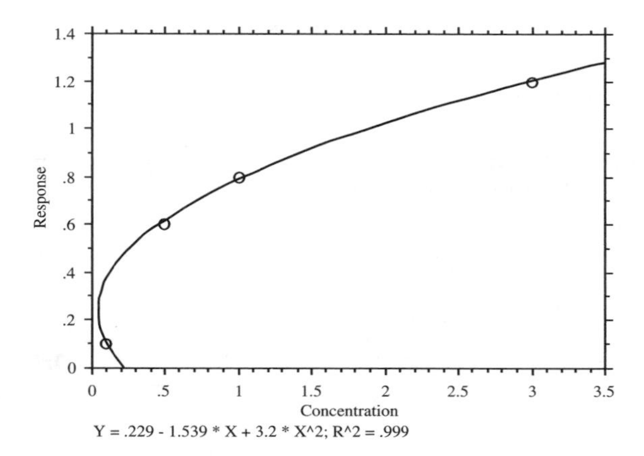

Figure 4-10. An unacceptable second degree calibration curve because it does not have unique response values for each concentration.

The third degree equations contain a term where the response is raised to the cubic power. The general equation for such a curve is,

$$y = ax^3 + bx^2 + cx + d$$

where y is again the concentration of the analyte, x is the response of the instrument, and a, b, c, and d are constants. The constant d is the intercept. If d is equal to zero, then the curve passes through the origin. Values of d other than zero mean the curve does not pass through the origin. Any four points can be connected using a third-degree equation. To demonstrate that a third-degree equation is suitable for a particular calibration, at least five different concentrations must be analyzed. Again, it is essential that the analyst visually examine the plot to verify that the equation is a suitable model.

External Standard Calibration

The external standard calibration consists of sequential injection of different concentrations of target analytes. As long as the injected volume of solution remains constant, and the operating conditions of the instrument are constant, then there is a relationship between the concentration of the target analyte in the standard and the integrated response from the instrument.

The simplest model is a linear equation through the origin. It assumes that there is a constant ratio between the area of the response and the concentration. One approach to this model is taken by calculating a calibration factor[1] (Cf) for each concentration level;

$$Cf = \frac{\text{Area TA}}{\text{Concentration TA}}$$

where TA is used to indicate the target analyte. The individual Cf are then used to calculate an average CF, which is used in the mathematical model as:

[1] The reader should be aware that the USEPA definition of the calibration factor and the later-discussed response factor is the inverse of the definition used by the major manufacturers of GC and HPLC software, Hewlett-Packard and Perkin-Elmer.

$$\text{Concentration TA} = \frac{\text{Area TA}}{\text{average Cf}}$$

where 1/Cf is equivalent to the slope, m, in the above general equation. The suitability of the model is determined through calculation of the percent relative standard deviation (%RSD) of the individual calibration factors. If the %RSD is less than a specified value, for instance 10%, then the model is considered to be valid.

Table 4-2. Calibration data for peak #1 of Arochlor 1016.

Amt ng/μL	Area	1/Cf
0.1	12116	8.25×10^{-6}
0.5	67664	7.39×10^{-6}
1.0	108984	9.18×10^{-6}
2.5	207675	1.20×10^{-5}
5.0	380720	1.31×10^{-5}
	Average 1/Cf	1.00×10^{-5}
	%RSD	24.8

An example of this process is presented in Table 4-2, for one of the indicator peaks in Arochlor 1016, analyzed on an ECD-equipped GC. The calculated %RSD for the individual calibration factors is 24.8%, a value indicating substantial deviation from the linear through the origin model.

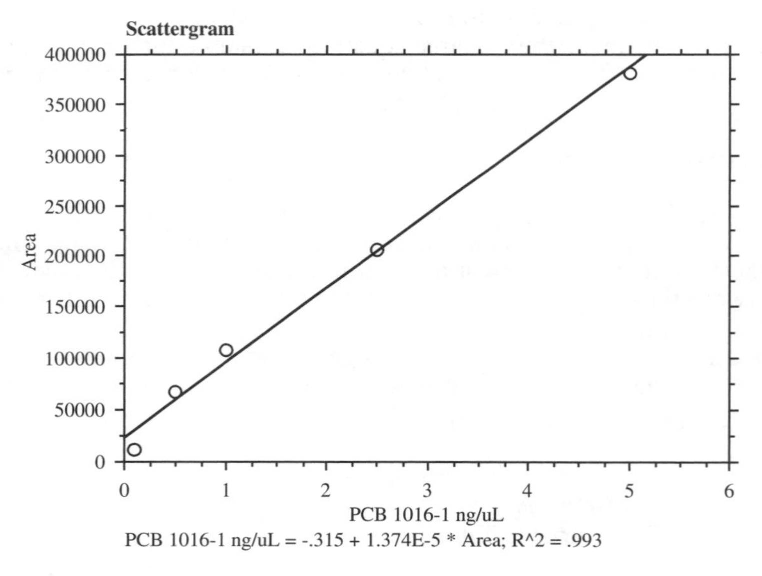

Figure 4-11. Plot of a linear not through the origin model for the calibration data in Table 4-2.

A better mathematical description of the calibration is obtained by not requiring the line to pass through the origin (zero). A regression is prepared, with the results illustrated in Figure 4-11. The overall fit of the line to the points is improved, and a reasonably acceptable correlation (r = 0.996) has been obtained. However, the equation does not accurately describe the behavior of the calibration at the low end. In particular, the lowest three points appear to describe a curve rather than the line. This type of behavior is very common in GC, especially on multipeak analytes such as the Arochlors.

There is an implication in the use of this calibration model in this situation that can have a serious effect on the client. Frequently analysis is performed in support of remediation efforts where the intent of the analysis is to demonstrate that all the contaminated soil has been completely removed. Action levels have historically been established for remediations. An action level is the concentration that indicates that no further removal is necessary. Recent decisions by state and federal regulators have been to base action levels upon risk evaluations, which in most cases will drastically reduce the concentration of the action level. Risk-based action levels are frequently at concentrations that are substantially below any analytically achievable detection limit. Negotiations between the engineering firm performing the remediation and the state regulator will adjust the action level to be equivalent to the detection limit of the laboratory.

This transfers the emphasis in the analysis to the detection limit, because anything above that level simply means more digging is required. If the laboratory assays a sample, and the response (say 13000) is above- but close to- the area counts of that exhibited by the lowest calibration point in Figure 4-11, the calculated result from the calibration equation is going to be lower than the established detection limit. This creates a Catch 22[2] where the laboratory has a hit for the target analyte above the response of the lowest calibration point, but the equation indicates no-hit. Obviously this is a situation that must be resolved.

The simple solution is to use an equation that more closely models the calibration data. A second degree regression-derived equation for the data in Table 4-2 is plotted in Figure 4-12. A slightly better correlation (r = 0.998) is obtained; however, the equation very closely models the behavior of the instrument at the important low end of the range. Unfortunately, the analyst now runs into additional regulatory guidance (EPA Method 8000B, SW-846) where to use a second degree or higher curve, either triplicate injections of 5 calibration levels or 10 individual calibration levels are suggested. This either doubles or triples the amount of time necessary to perform the calibration. Another option is to reduce the range of the calibration. In this example that might be analysis of 5 calibration standards covering the range 0.1 to 1.0 ng/µL.

---

[2] Heller, J., 1961. *Catch 22*. Simon & Schuster, New York, NY (A humorous fictional account of requirements that are mandated by regulations but unsupported by actual observation)

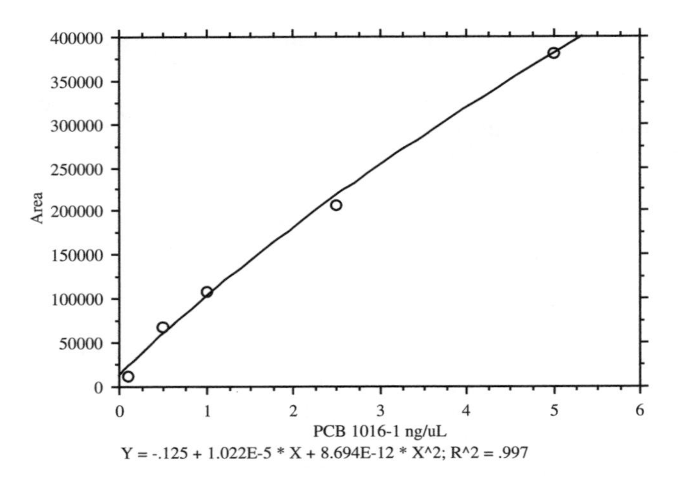

Figure 4-12. Plot of second degree model for calibration data in Table 4-2.

There is a suitable middle ground that only requires addition of two more calibration standards and still maintains the 0.1 to 5.0 range. These additional standards (0.3 ng/µL and 0.75 ng/µL) are included in Table 4-3.

Table 4-3. Calibration data for peak #1 of Arochlor 1016 (Table 4-2) with two additional points.

Amt ng/µL	Area	1/Cf
0.1	12116	8.25×10^{-6}
0.3	39890	-
0.5	67664	7.39×10^{-6}
0.75	88324	-
1.0	108984	9.18×10^{-6}
2.5	207675	1.20×10^{-5}
5.0	380720	1.31×10^{-5}

The next step is to set-up two separate 5-point linear calibrations for the analysis. These are indicated in Figures 4-13 and 4-14. The analyst then picks which range and corresponding equation gives the best data for the sample. A second-degree regression equation for all seven points is presented in Figure 4-15.

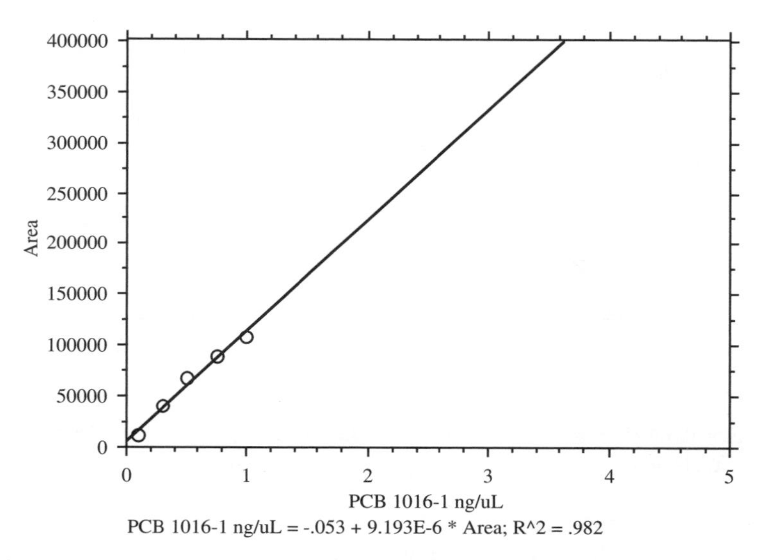

Figure 4-13. Low-level 5-point linear regression equation for Arochlor 1016-1 calibration data in Table 4-3.

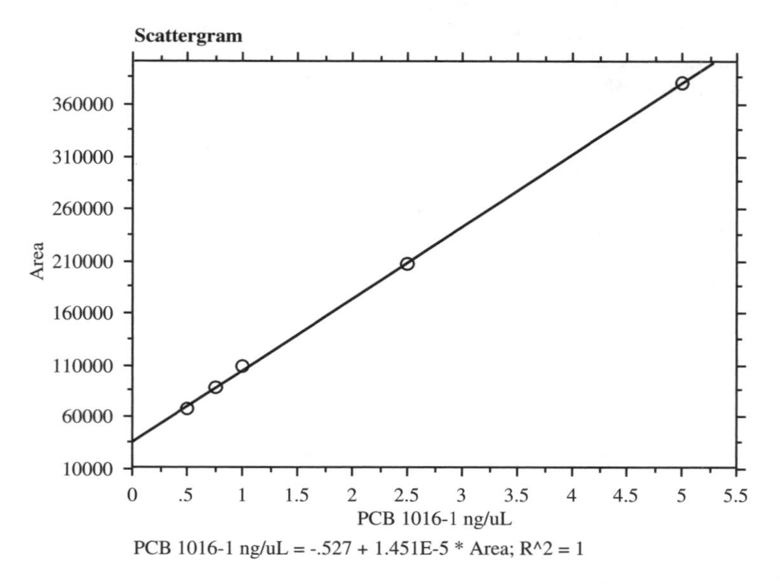

Figure 4-14. High-level 5-point linear regression equation for Arochlor 1016-1 calibration data in Table 4-3.

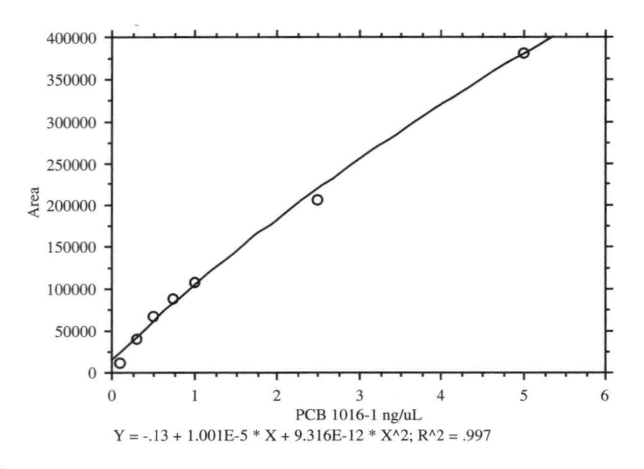

Figure 4-15. Second-degree regression equation for all seven points in Table 4-3.

Internal Standard Calibrations

The internal standard calibration evaluates the response of the target analyte in comparison to the response of a known amount of an added compound. The added compound, the internal standard, is present in the same amount in all calibration solutions and samples. The intention behind use of the internal standard is to minimize effects of injection volume and other chromatographic and detector variations. The assumption is that any variation in the response of the target compound is also reflected in the response of the internal standard.

A model for the calibration is presented as a schematic in Figure 4-16. The various calibration levels of the target analyte are shown as the nested peaks on the right, while the invariant internal standard is indicated by the single peak on the left.

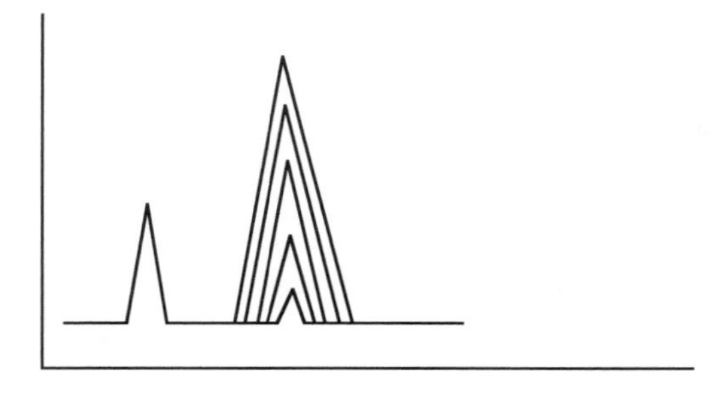

Figure 4-16. Schematic of an internal standard calibration.

The simplest expression of the use of the internal standard calibration is that of the response factor (Rf). The response factor is calculated for each level of calibration, then the average of the individual response factors is used for target analyte quantitation. The percent relative standard deviation (%RSD) of the set of response factors is a measure of the linearity of the calibration.

$$Rf = \frac{\text{Area TA x Conc IS}}{\text{Area IS x Conc TA}}$$

The slope (m) of the linear through the origin calibration equation, y = mx, is a constant composed of the concentration of the internal standard divided by the response factor. The instrument response (x) is the ratio of the area of the target analyte divided by the area of the internal standard.

$$\text{slope} = \frac{\text{Conc IS}}{\text{Rf}}$$

For certain combinations of target analyte and internal standard, such as phenanthrene and phenanthrene-d_{10}, the linear through the origin model works very well and a low %RSD (<5%) is obtained. Current guidance (EPA Methods 8260B and 8270C) allow the %RSD to rise as high as 15% for assumption of linear through the origin behavior for other calibrated compounds. Well-maintained instruments will achieve this criteria for most compounds.

For some combinations of internal standard and target analyte, the relative response of the areas is actually not linear with concentration, and more advanced equations should be used. One example, found in EPA Methods 624 and 625, that date back over 20 years, sets up an equation between the response factor and the ratio of analyte and internal standard areas. More modern approaches directly calibrate the amount of analyte in terms of the ratios of analyte and internal standard area. Most instruments offer a menu of calibration options, and finding and choosing the most suitable equation is little more than a simple mouse click. Again it has to be stressed that the analyst must visually examine the graph of the calibration to insure that an appropriate model has been selected. The instrument software cannot perform this evaluation.

Internal standard calibrations work best when the internal standard is very similar to the target analyte. The preferred case is when the internal standard is an isotopically substituted analog of the target analyte, such as the above phenathrene example. The substitution of deuterium for hydrogen is the most common and the cheapest; however, other substitutions are seen such as carbon-13 or chlorine enrichment. The deuterium substituted internal standards will have all the hydrogens in the molecule replaced by deuterium, and the percent of molecules in the purchased material that exhibit the proper substitution will be close to 100%. There is a vanishingly small probability of encountering a perdeuterated compound in an environmental sample. When carbon or another element substituted materials are used as internal standards, generally there is only one specific atom in the molecule that is subject to substitution, and within the manufactured material the degree of successful substitution is often not 100%; 80 to 95% being common. Depending on the exact element that is isotopically enriched, a correction factor may need to be added to the calculation to account for the less than 100% enrichment. And, there may be a percent of the naturally encountered target analyte that exhibits the substitution. Examples of these correction factors are presented in EPA Methods 1665 and 1666.[3]

In the majority of uses of the internal standard calibration, the internal standard is chemically different from the target analytes. In EPA Method 8270C, the internal standards are perdeuterated polyaromatic hydrocarbons, while the target analytes included in the Method include a wide range of chemical types such as esters, nitrosoamines, phenols, and chlorinated pesticides, in addition to the PAH. In cases where the internal standard is chemically different from the target analyte, they too

[3] USEPA, *Analytical Methods for the Determination of Pollutants in Pharmaceutical Manufacturing Industry Wastewater* (EPA-821-94-001) February, 1995

should at least elute close in time to each other. For example, an internal standard calibration where the internal standard elutes at 5 minutes into the GC run, and the target analyte elutes at 25 minutes, is not going to be as robust a technique as one in which the internal standard elutes at 23 minutes. When a wide range of elution times are exhibited by the target analytes in a method, several internal standards that collectively exhibit elution times covering the range should be used.

The internal standard solution is normally added to the sample extract immediately before performing the injection of the extract into the instrument. EPA Method 525.2; however, is one example of a procedure where perdeuterated PAH internal standards are added to the sample prior to the sample preparation. Isotopic dilution is a variation on the use of internal standards, where the standard solution is added to the sample prior to extraction or preparation steps. The internal standards in this technique normally include an isotopically labeled analog for each of the target analytes. This calibration technique will minimize analyte recovery variations due to matrix complexities and sample manipulation.

Initial and Continuing Calibration Verification

Most analysts do not prepare their own calibration standards to the extent that a neat solid or liquid of each target analyte is maintained on-site and is weighed out each time a new set of calibration standards are needed. Instead multi-component stock solutions are purchased from vendors, and the stock solution is serially diluted to make the needed set of standards. Sometimes the vendor will make a mistake in the preparation of the stock solution, such as addition of the wrong compound[4], or adding an incorrect amount. These errors in the concentration of the stock solutions will have a drastic effect on the reliability of sample data. However, this may not be apparent during the calibration procedure. Because of the potential for these type of errors to occur, the prudent analyst should purchase or prepare a duplicate of one of the calibration solutions from a second vendor that is unassociated with the vendor of the primary standards. Analysis of this second-source standard will verify the accuracy of the primary standards and help to catch these mistakes.

GC and HPLC instruments do not maintain calibrations forever. The length of time that an instrument will retain a calibration depends on the sample matrix/sample extract introduction technique and the individual detector used on the instrument. Instruments that use a purge and trap type sample introduction will, in general, retain calibrations a lot longer than those instruments that rely on direct sample injection, all other factors being constant. The mass spectrometer response will degrade over time as ionization residue builds up inside the source and focusing lenses. Some detectors, for example, the electron capture detector, can be characterized as having a constant detector response that is a transitory phenomenon at best, and the instrument requires at least daily calibration. On the other hand, the amount of time that is required for a full calibration of an instrument is substantial, often taking the better part of an entire work day.

For these reasons, it is essential to regularly check that the instrument maintains calibration. This is performed by re-analysis of a calibration solution and insuring that the response is within certain criteria for the calibration verification. Often a deviation of less than 20% from the response obtained during the initial calibration is used as a success indicator. The calibration check must be performed at least once each day of

---

[4] Smith, R.-K., 1998. Benzidine? Really? *Proceedings of the 14th Annual Waste Testing and Quality Assurance Symposium*, 13-15 July, 1998, Crystal Gateway Marriott, Arlington VA., pp 40-43.

sample analysis, and for techniques using two-dimensional detectors a re-check after every 10 samples is normal.

When an analyte does not pass calibration verification criteria, the draconian viewpoint is that the instrument is no longer calibrated for the compound and that all data for that analyte are suspect and must be rejected. As in most areas of environmental analysis, the actual situation is one of shades of gray. If the calibration verification is only slightly outside the normal allowances the data may still be reliable, and in most instances the data are still usable. The questions to be asked are; How far out of calibration is the analyte? In what direction? And how many analytes are affected? Let's look at the various combinations.

In the cases where the analyte is responding high (biased high), in other words giving a larger result than the true value of the calibration verification, very rarely will data have to be rejected. Since the analyte results are assumed to be greater than the true amount in the sample extract, there is no problem with detecting and reporting the presence of the analytes. Thus all positive sample results can be evaluated as estimates, while all non-detects are reliable. A high bias on a non-detected analyte is still non-detected.

The situation is different in cases where the analyte is responding low (biased low), in other words giving a smaller result than the true value of the calibration verification standard. It is really necessary to try to find the root cause of the low response before data can be realistically evaluated.

The most common cause of low responses, particularly when a number of the target analytes are low[5], can be traced to the internal standard solution. As internal standard solutions are used they tend to lose solvent through evaporation. This results in a real increase in concentration. Addition of a constant amount of solution over several days gives an actual increase in mass of internal standard added. Since the amount of internal standard is inversely related to the calculated amount of target analyte, the end result is an apparent decrease in target analyte concentration. The same effect can be seen with a one-time mistake in volume of added internal standard solution.

The first response of the instrument operator in these cases should be to re-prepare and re-analyze the calibration verification standard. However, when this corrective action is not performed, and samples are analyzed under these low response conditions, the reviewer should be very careful during evaluation. If an analyte is reported present, it is a safe assumption that the quantitative result is biased low; however, there is no reason to believe that the identification is faulty. The proper action is to flag the results as estimates if they are not already flagged.

It is the analytes that are reported as not-detected or below a quantitation or detection level that present the problem. If the internal standard is the root cause, then there is no impairment in the instrument's ability to generate a response to the analyte and thus identify the analyte as present or not-present. It is the quantitative value assigned to this detection capability that is affected. If the analyte is present in the sample the instrument will correctly respond to it; however, the generated value may be below some applied reporting limit. It is not appropriate to report that the analyte was not detected in the sample, and the result in such a case is unreliable. The laboratory should report that the analyte was detected but that the quantitation limit is an estimate. However, when the instrument detects no trace of the analyte, there really is no confirmed presence in the sample. Thus the cases to distinguish are: 1) analyte present in the sample but artificially below a reportable cut-off; and 2) not detected in

---

[5] By "number of analytes low" means almost all the analytes are reported on the low side of the true value: most may still be within the acceptance percentage, and some outside the range, but overall the whole QC measure indicates a low recovery.

the sample. The laboratory can help guide the data evaluator by providing a succinct discussion of the situation in a Case Narrative. But, the sample results should never have left the analytical area of the laboratory under this scenario.

The other causes of low calibration verification results stem from degraded ability of the instrument to respond to the analyte. Some of the more sensitive analytes can be destroyed by dirty injectors, high injector temperatures, reactive residues in the column, reactive sites in the column liquid layer, and dirty detectors. Examples of some of these sensitive analytes include DDT, endrins, benzidines, chloroethylvinyl ether, anilines, and phenols. At the same time these materials are exhibiting calibration problems, the less reactive analytes such as the PAH and PCB will continue to demonstrate acceptable calibration verification responses. Under these conditions it is impossible to determine whether the affected analyte was originally present in the sample extract or not. Analytes reported as present are verifiably present, but the quantitation is an estimate. Analytes reported as not-detected are unreliable. Analytes reported as below a quantitation level might be present, particularly when the quantitation level is significantly higher than instrument capability. Again the Case Narrative can be used to help clarify the exact situation.

Boundaries of Calibration

Each calibration has boundaries. These boundaries are determined by the capability of the instrument, the so-called response range. Part of the job of the instrument operator is to establish the boundaries for each instrument. Different calibration techniques can change the boundaries, and the operator should be aware of these differences. The two components of the boundaries are the lowest calibrated level and the maximum calibration level. Separate techniques are used for these determinations.

The lowest calibrated level is determined from several considerations. The first is the innate ability of the instrument to produce a signal that is distinguishable from the background noise of the instrument. Each instrument produces noise. Some elements that contribute to the noise include random electronic signals from the power source, the heated zones of the instrument, and the data processing components, column bleed from the separation column, and standing backgrounds from the detector. There are others. The intensity of these sources is not constant over time, and one of the intentions of preventative maintenance is to minimize these sources of background noise. It is noteworthy that these contributors to the background can never be entirely eliminated.

A technique for determining the instrument noise level is the Instrument Detection Limit (IDL) study, analogous to the IUPAC critical level. Frequently encountered in the instrumental analysis of metals by inductively coupled plasma atomic emission spectrometry (ICP-AES) and graphite furnace atomic absorption (GFAA), IDL studies are not commonly performed in organics analysis. The technique of an IDL study is to measure the level of background noise from the instrument. Based on this average background a quantitative threshold is established. The threshold is set above the average noise level (mean) at three times the standard deviation of the noise. The interpretation of this threshold is that signals below the IDL can not be distinguished from instrument noise, while signals above the IDL may be distinguishable from noise. A visual representation of the IDL concept is presented in Figure 4-17.

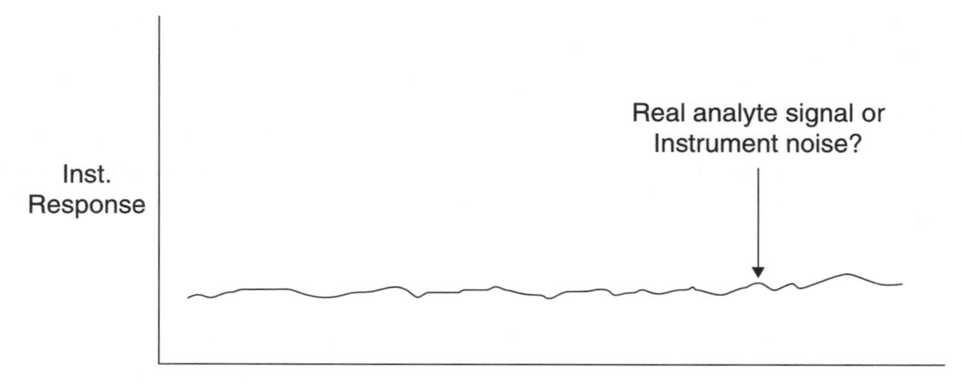

Figure 4-17. Representation of an IDL in a chromatographic analysis.

The intent of the IDL is to prevent false positive results from instrument artifact signals. However, the Method Detection Limit (MDL) study is not designed to prevent false negative results. It is the concentration of analyte in the sample that will generate a positive signal 99.7% of the time. Several characteristics of the MDL should be stressed. First, in most MDL studies, a sample of concentration equivalent to the MDL value is not analyzed. Generally in organics analysis, the tested concentration is from 2 to 10 times the value of the MDL with the higher multipliers being encountered most frequently. Second, the MDL is not a study of quantitative accuracy; instead it is a measure of precision. Third, the MDL study is conducted on ideal samples prepared from reagent grade water. In other words, there are no matrix interferences or co-extractables in the samples. The MDL is only applicable to samples that have a similar freedom from interferences. Within the environmental realm the closest type of sample to these characteristics are drinking waters, even though they to at times may contain miscellaneous organic matter. A visual representation of an MDL study for an organic analyte by a chromatographic technique is presented in Figure 4-18. Note the essentially flat and well-behaved baseline.

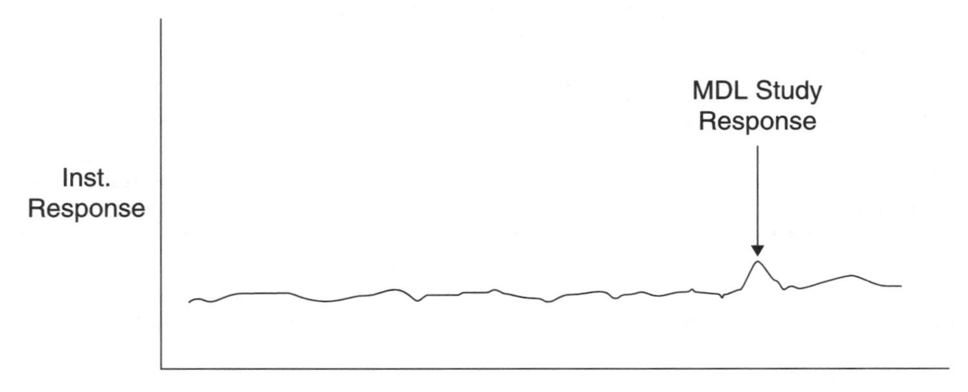

Figure 4-18. Representation of an MDL study result.

When the normal sample is analyzed, it will frequently exhibit background signals. Figure 4-19 illustrates the MDL picture of Figure 4-18 superimposed over a representative sample background.

The key lesson is that an MDL can not be an expected quantitation level. The quantitative inaccuracy for analysis of samples that contain the target analyte at the MDL is estimated to be at least 100%.[6] Many laboratories and most consultants and regulators interpret the MDL as a quantitation limit when it was never designed nor intended to function as such. The only correct interpretation of the MDL is that if the sample contains the analyte at that concentration, a positive signal will be generated by the instrument regardless of whether the signal is lost in the sample background or not. The MDL is a qualitative concept; definitely not a quantitative one.

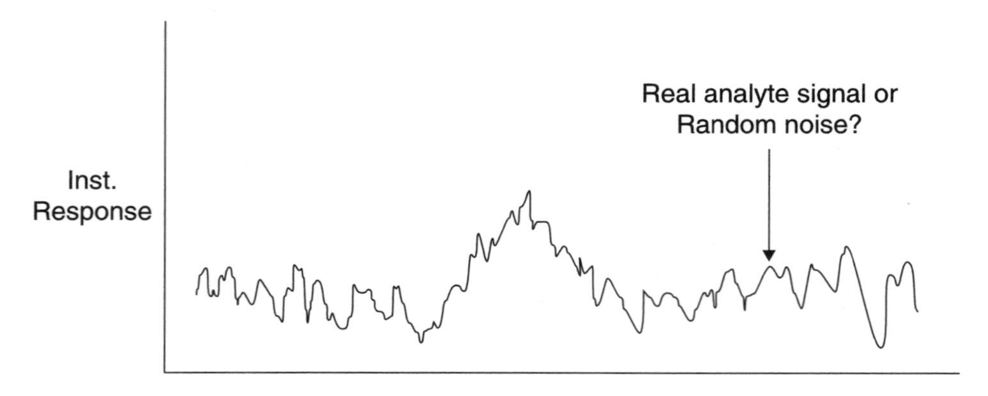

Figure 4-19. Representation of an MDL value superimposed on a real sample extract.

Reliable quantitation of analytes in samples can only begin at some higher concentration than the MDL. Current thought about the lowest reliable quantitation limit is that a multiplier of from 3 to 5 times the MDL value generates a usable value for well behaved samples.[7] For samples that still exhibit considerable interferences, even after application of suitable clean-up procedures, the lowest reliable quantitation level can be still higher.

For purposes of verifiable and defensible calibrations, these considerations indicate that the concentration of the lowest calibration standard should be at the quantitation limit; in other words 3 to 5 times the MDL.

Recently more and more states' regulatory limits for clean-up have been based on risk assessments. Risk assessments take into account the lowest observable effect concentration of a compound upon a test organism (generally not a human or primate, and sometimes not even a mammal) and translate that concentration to the increased weight of the average human. Then an acceptable risk factor is applied to the concentration. The risk factor is often in the range of 1/1000 to 1/1000000, depending on the projected health effect of the compound and sometimes the fate/transport of the compound in the environment. The risk-based concentration frequently is calculated to fall in to a range that is substantially below any demonstratedly achievable laboratory

---

[6] Solsky, J., 1998. Questionable practices in the Organic Laboratory. *Proceedings of the 14th Annual Waste Testing and Quality Assurance Symposium,* 13-15 July, 1998, Crystal Gateway Marriott, Arlington VA.; Solsky, J., 1999. Questionable practices in the Organic Laboratory: Part II. *Proceedings of the 15th Annual Waste Testing and Quality Assurance Symposium,* 18-22 July, 1999, Crystal Gateway Marriott, Arlington VA.

[7] FL DEP QA Section; U.S. Army Corps of Engineers, October, 1998. Shell for Analytical Chemistry Requirements, Omaha, NE.

quantitation limit. By “demonstratedly achievable”, I mean that IDL and MDL studies have been performed for each analyte; a quantitation limit based on the MDL study results has been established; and the lowest point on the calibration curve is equal in concentration to the quantitation limit.

State and federal regulators will often listen to reasonably convincing arguments about analyte quantitation and detection limits when those arguments are supported by IDL, MDL, and calibration data. Achievable clean-up levels can then be negotiated, unless the regulator has had the prior experience of some XYZ laboratory reporting “not detected” at whatever level the risk-based assessment generated, without any regard to MDL studies or comparable lowest calibrated standard. Unfortunately there are all too many laboratories, who instead of defending scientific vigor, take the easy way out and simply falsify unsupported data (i.e. reporting non-detected when there are no supporting IDL, MDL, or other instrument/method sensitivity determinations). In these situations the response from the regulator is frequently, “XYZ laboratory reported results to these levels so why don’t you, as an engineering firm, use them, instead of listening to these excuses of why the laboratory can’t report to these levels.” It’s a clear indication that the regulator has accepted unreliable data in the past. Indefensible data and inexcusable laboratory practices are perpetuated through events just like these.

All instruments will exhibit detector saturation at some upper level of analyte concentration. Saturation is where there is no longer a regular change in detector response with increased concentration of analyte. The maximum calibration level must be defined for each analyte and for each instrument. Some detectors such as the PID and FID can routinely be calibrated almost up to the saturation level, while others, such as the ECD and MS, suffer incremental loss in sensitivity after exposure to saturation levels of analytes and can be very slow to recover to baseline conditions.

The maximum level will differ depending on the mathematical description of the calibration. Linear calibration models are very sensitive to instrument saturation, while second- and third- degree equations can tolerate at least modest beginnings of saturation.[8] It is in the best interests of the analyst and laboratory to evaluate maximum calibration levels and to calibrate each instrument with as wide a range of analyte concentrations as possible. Reasons behind this are in terms of analytical reliability, time, and effort.

It is very poor analytical practice to report results of sample analysis that are determined by extrapolation of calibration curves. The practice might be scientifically defensible if the analyst has performed an instrument saturation study and knows exactly how far the calibration model can be extended before significant deviations from the model are observed. Although an obvious study for an operator to perform, in reality it very rarely is. In the general case, the rule is never to report an analytical result that falls above the maximum concentration of the calibration standards. The common solution in the event of responses that exceed the calibration is to perform a quantitative dilution of the sample extract, and re-analyze it so that a response is obtained within the bounds of the calibration. The calculated result is then multiplied by the dilution factor.

Determination and use of the maximum calibration range for the instrument is a time saver. The savings are realized in reducing the number of dilutions and re-analyses that must be performed to generate usable results within the bounds of the calibration curve.

---

[8] If a second- or third- degree equation is being used to model beginning saturation, the analyst should regularly check the validity of the description by analysis of a continuing calibration standard that has a concentration that is in this upper level. The point of saturation onset can change depending on the age and condition of the detector.

C. Visual Evaluation of Calibrations

Most published methods will have numerical criteria for evaluation of calibration data. Some examples of these types of criteria have been presented above, such as a maximum allowance on a %RSD of a set of individual calibration or response factors, a maximum allowable percent deviation (%D) on a calibration verification, etc.

Lists of numbers representing calibrations generally are incomprehensible to most analysts and data reviewers. In all situations it is preferable to have a picture or graph of the calibration points complete with the track of the calibration equation through the points.

D. Product Analysis

As an aside, there is a substantial difference between trace analysis and product verification. Product verification is a quality control procedure performed by a manufacturer of a chemical formulation. Manufacturers will often sub-contract this aspect of their operations, and a frequent recipient of these sub-contracts are environmental laboratories that are accustomed to trace level analysis. In trace level analysis an uncertainty in the results of up to a factor of 2 (-50% to + 100%) from the true value of the analyte in the sample is commonly inconsequential as far as remediation and enforcement decisions are concerned. A deviation of ±20% in the calibration is within the expected uncertainty of the analysis.

This level of uncertainty does not exist in formulation verification analysis and is completely unacceptable. When a formulation is stated as 5.0% concentration often the allowable absolute bounds on this concentration are on the order of ±0.2% (4.8 - 5.2%), and sometimes even tighter. This translates to an acceptable relative uncertainty in the analysis of 4%. A reported analytical result of 5.0% when the label states 6.0% is going to trigger major events on the part of the manufacturer, including temporary shutdown of production until the process aberration is found and corrected. Lawsuits between compounders and raw material suppliers can be started as a result of the analytical results. If the formulation contains a registered pesticide, an analytical result of 7.0% when the label states 6.0% can initiate regulatory action.

It should be obvious that an allowance of ±20% on a calibration verification is going to significantly exceed the analytical requirements for product verifications. Other common habits of persons performing trace analytical procedures, i.e. use of microsyringes to perform dilutions, eyeballing volumes, and approximate weight determinations (1.0 g instead of 1.000 g), are completely unacceptable in product verification.

CHAPTER 5

Quality Control Acceptance Limits[1]

The evaluation of laboratory data looks at, among other things, a number of quality controls. These include calibration, calibration verification, blanks, duplicates, matrix spikes, laboratory control samples, performance evaluation samples, instrument performance, and surrogates, depending on exactly which parameter has been analyzed. Each quality control has something to say about the analysis, and the information is relayed to the data evaluator in terms of a number. In previous chapters the idea of examining the raw data visually and trying to understand how the QC results are applicable to the sample results has been stressed, but in practice the majority of quality control evaluations are simply based on number comparisons. If the QC measures meet the numerical criteria everything is good, while failure to meet the criteria equates to sample analysis failure, with the word "failure" always emphasized. For a calibration the number may be a regression coefficient (r) or percent relative standard deviation (%RSD). The success of a calibration verification is determined as a percent difference (%D). Duplicates, matrix spikes, laboratory control samples, and surrogates generate numbers for precision as relative percent difference (RPD) and for accuracy as percent recovery (%R).

The first action of the data evaluator who thinks in these terms is to take the number, examine it against the normal range of the quality control, and then make the decision as to whether the quality control result is passing or failing. For some of the quality controls the decision point is a single number. For example, when looking at blank results, the detection or reporting limit of the test is the acceptance limit. A blank result below the reporting limit is good, while a blank result above the detection limit may indicate that the laboratory has worrisome levels of ambient contamination, and all data need to be very closely examined. A continuing calibration is satisfactory if the %D is less than 20%. However, for evaluations of precision and accuracy, the results are compared to acceptance ranges, such as 0-20 for relative percent difference or 85-115 for percent recovery.

This chapter examines acceptance ranges - where they come from, what they mean, and how to use them to the maximum effect. It also presents the pitfalls inherent in by-the-number QC evaluation.

The story begins with control charts. There are a number of different types of control charts; however, the simplest is the Shewhart Chart, developed in the 1920's by Dr. Walter A. Shewhart of Bell Laboratories. These consist of a piece of square-ruled graph paper. The metric of the quality control, such as %R, is labeled on the vertical axis, and the date of the quality control is placed on the horizontal axis. Each day the result of the quality control for the test is plotted. Figure 5-1 presents a Shewhart Chart for recovery.[2]

---

[1] An abridged version of this chapter was published in The Georgia *Operator*, Georgia Water and Pollution Control Association, Atlanta, GA, 36(2):48, 50, 58, 62 Spring, 1999.

[2] Smith, R.-K., 1999. *Handbook of Environmental Analysis, 4th Edition*, Genium Publishing, Schenectady, NY 1-800-243-6486

The preparation and maintenance of control charts is a mandatory quality control[3] for most test procedures used to generate data for regulatory reporting under the drinking water and wastewater programs.[4] In the past the control charts were maintained on paper, and many analysts couldn't be bothered with regularly updating the charts and re-calculating control and warning limits. With the advent and widespread use of computers and Laboratory Information Management Systems (LIMS) for quality control data collection, these required data are always present and easily manipulated to generate the needed acceptance ranges. There is no longer any excuse for a laboratory not to have current in-house acceptance limits for each analyte in each test.

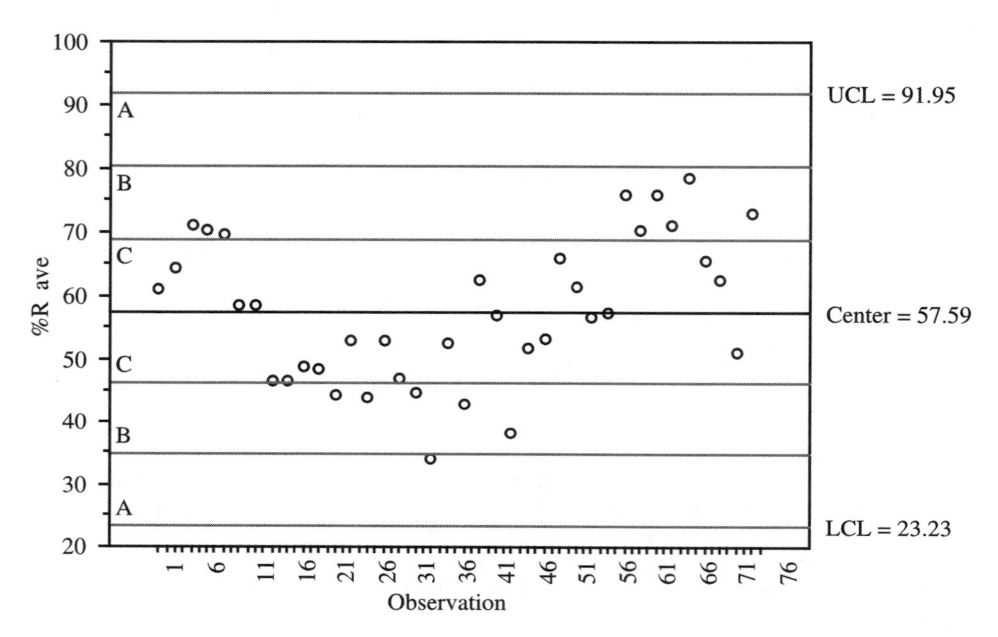

Figure 5-1. Shewhart control chart for accuracy, %R.

When at least 20 data points have been plotted on the control chart, the control limits can be calculated from the average and standard deviation of the results. In Figure 5-1, the average of the points is 57.59, and a line is drawn across the chart at this position. Six other dotted lines are present in the Figure. These are at ± 1 standard deviation (zone C), ± 2 standard deviations (zone B) and ± 3 standard deviations (zone A). The latter lines are additionally labeled as UCL (upper control limit) and LCL (lower control limit).

Standard deviation is a statistical description of data. Statistics means that a small number of data points are used to predict the behavior of either a much larger data set or future events. In the case of control charts, the intention is the prediction of future

[3] USEPA, 1979. *Handbook for Analytical Quality Control in Water and Wastewater*, EPA-600/4-79-019, NTIS PB-297451; USEPA, March, 1997. *Manual for the Certification of Laboratories Analyzing Drinking Water, Fourth Edition*, EPA 815-B-97-001, available on the Internet from the EPA website, www.epa.gov/ogwdw

[4] Smith, R.-K., 1999. *Lectures in Wastewater Analysis and Interpretation*, Genium Publishing, Schenectady, NY 1-800-243-6486.

events, i.e. the next time a recovery is performed. The common interpretation is that 67% of the future observations should fall within mean ± 1 standard deviation (zone C), 95% within mean ±2 standard deviations (zone C + B, warning limits), and 99.7% within mean ± 3 standard deviations (UCL to LCL).

There are several conditions of the data set for the statistics to be valid and usefully predictive. Unfortunately in most laboratory operations some of these conditions are not met. The first condition is that the data must be randomly distributed. It should be obvious from Figure 5-1 that these data are not randomly distributed, there is a distinct V-shape to the points. The statistically proper 28:67 distribution is approximately met when comparing zones B and C; however, the other Western Electric rules (nine consecutive points on one side of the centerline, six consecutive points in the same trend, etc.) are regularly violated in laboratory data[5], without being indicative of serious process control problems. Reasons behind these observations are many. A finite lifetime of solutions and reagents is cited as one reason. Required periodic instrument maintenance, adjustment, and consumable part replacements are others. The innate variability of the samples being analyzed is still another. These events are a normal part of the analytical process and must be taken into account in the overall evaluation of the system. On the other hand, during the processing of laboratory samples, analysts frequently run into intractable samples (giving no recovery of the matrix spike or surrogate) and what are best described as laboratory blunders[6] (double spiking a sample, or spilling half of it). Inclusion of these data points in statistical evaluations can widen the UCL and LCL to totally useless values.

Laboratories must use an outlier identification mechanism to void these data points and then use the reduced data sets to generate useful limits. The common outlier identification procedures are the Grubbs Test[7], the Dixon test[8], Rosner's test[9], and Walsh's test.[10] They have different applicabilities, depending on the size of the data set.

The Dixon test (also called the Extreme Value test) is applicable to small data sets. The size of the data set determines the form of the Dixon test. First rank all the data points according to size from smallest to largest. If the data set contains seven or less points, next compute the ratio of the separation between the possible outlier and the next nearest point divided by the range (largest point - smallest point). Compare the ratio to the 95% critical value: 3 points - 0.941, 4 points - 0.765, 5 points - 0.642, 6 points - 0.560, 7 points - 0.507. If the data set contains from 8 to 10 points a reduced range is used as the divisor, i.e. drop the farthest away point. The 95% critical values are: 8 points - 0.554, 9 points - 0.512, 10 points - 0.477.

---

[5] Berger, W., H. McCarty, R.-K. Smith, 1996. *Environmental Laboratory Data Evaluation*, Genium Publishing, Schenectady, NY, 1-800-243-6486.

[6] Make no mistake, all laboratories on a regular basis make mistakes, break bottles, switch samples, and a myriad of other SNAFU's. The lab that never makes a mistake and never informs the client about it has some serious ethical problems in addition to the normal suite of lab haphazards.

[7] Taylor, J.K.,1987. *Quality Assurance of Chemical Measurements*, Lewis Publishers, Chelsea MI; Grubbs, F.E., and G. Beck, 1972. Extension of sample sizes and percentage points for significance tests of outlying observations. *Technometrics* 14(4):847-854.

[8] Taylor, J.K.,1987. *Quality Assurance of Chemical Measurements*, Lewis Publishers, Chelsea MI; Dixon, W.J., 1953. Processing data outliers, *Biometrics* 9(1):74-89.

[9] Rosner, B., 1975. On the detection of many outliers. *Technometrics* 17:221-227.

[10] Walsh, J.E. 1958. Large sample nonparametric rejection of outlying observations. *Annals of the Institute of Statistical Mathematics*, 10:223-232.

Example

7 item ranked data set: $N_1, N_2, N_3, N_4, N_5, N_6, N_7$

If N_1 is a suspected outlier:

$$\text{Dixon ratio} = \frac{N_2 - N_1}{N_7 - N_1}$$

At a 95% confidence level if the Dixon ratio is greater than 0.507 then the point is an outlier.

9 item ranked data set: $N_1, N_2, N_3, N_4, N_5, N_6, N_7, N_8, N_9$

If N_9 is a suspected outlier:

$$\text{Dixon ratio} = \frac{N_9 - N_8}{N_9 - N_2}$$

At a 95% confidence level, if the Dixon ratio is greater than 0.512 then the point is an outlier.

For the simple seven or less items in the set case the Dixon test is equivalent to the Q test.[11] The Dixon test has other confidence levels and Taylor's book should be consulted for a complete discussion.

The Grubbs test (also called the Discordance test) is applicable to large data sets and makes use of the standard deviation and the mean of the entire data set. The Grubbs ratio is calculated from the separation between the suspected outlier and the mean, divided by the standard deviation. The 99% confidence critical values are: 10 points - 2.410, 15 points - 2.705, 20 points - 2.884, 25 points - 3.009, 50 points - 3.336, 100 points - 3.600.

The Grubbs test has critical values for other confidence levels, such as 95% and 90%. These are not very useful when evaluating data sets to generate warning (95%) and control limits (99.7%) as you may be deleting points that reflect normal operating conditions. The 99% critical values serve to delete the outlandish points without severely affecting the normal mean and distribution of the random error events.

Rosner's test and Walsh's test are used to detect multiple outliers in large data sets. Rosner's test can handle up to 10 outliers in sample populations that contain at least 25 members. Walsh's test at the 95% confidence level is applicable to data sets of at least 220 members, but can handle both large and small value outliers, and the number of outliers tested is open-ended. Rosner's test is tedious since it requires iterative calculations and comparisons. Walsh's test is easier to apply, requiring the calculation of four constants, then testing the outliers as a group with the constants. It is ideally suited for the large data sets that are part and parcel of acceptance limit determinations.

Example of Walsh's test

Order the data from the smallest to the largest; $X_1, X_2, \ldots X_n$
Identify the number of possible outliers, *r*. *r* can equal 1.
Compute the following constants:

$c = \sqrt{2n}$ rounded up to the next integer, for instance 3.31 becomes 4

[11] Hargis, L.G., 1988. *Analytical Chemistry Principles and Techniques*, Prentice-Hall, Englewood Cliffs, NJ

$k = c + r$

$b^2 = 1/\alpha$, where $\alpha = 0.05$ if $n > 220$

$$a = \frac{1 + b\sqrt{[c - b^2]/[c - 1]}}{c - b^2 - 1}$$

then:

the r smallest points are outliers if:

$X_r - (1 + a)X_{(r+1)} + aX_{(k)} < 0$

the r largest points are outliers if:

$X_{(n+1-r)} - (1 + a)X_{(n-r)} + aX_{(n+1-k)} > 0$

Both Rosner's and Walsh's tests are described in an EPA guidance manual[12] and are contained in a software package available from EPA[13] along with Dixon's and Grubbs' test.

Well, all that was pretty dry and relatively unexciting. But if you have persevered this far, let's go on to talk of acceptance limits.

The reason that it was necessary to discuss control charts is that the warning and control limits are exported from control charts to quality control summaries and are what we term acceptance limits. Under no circumstances should acceptance limits ever be interpreted as stone walls. Under this paradigm, results inside the walls are good, while results outside the walls are garbage, and any associated data are worthless. Nothing could be farther from reality. However, this is the common interpretation used by many data users and all computers.[14]

Statistics are gray numbers, not black and white. The most useful and most frequent choice for acceptance limits are the ±2 standard deviation lines, the warning limits. Acceptance limits constructed from the warning limits give a 95% predictive interval. This means that 95% of future results should fall within the acceptance limits, for example 64-103 for percent recovery. It also means that 5% of the results should fall outside the interval. Or in normal batchwise analysis, one result out of every 20 is expected to fall outside the acceptance range due to unavoidable random events.

Acceptance limits can also be based upon control limits. These are ±3 standard deviations from the mean and are frequently interpreted as the 99.7% confidence intervals. When there are no other influences upon the result other than random variations, then only three out of 1000 data points would lie outside the control limits. Unfortunately real-world samples are analyzed in environmental labs, and these samples often present non-random complications to the analysis.

Where data users can elevate themselves above the black and white limitations of the computer is when they start asking, "Well the recovery is outside the acceptance range. How far out is it?" and "Are all the points out or is it just this one?" and "What do these quality control results tell me about the data reported on my sample?" These questions are evidence that judgment is being exercised. A single point in a batch slightly outside the warning range is probably the 1 in 20, particularly when it lies within the control range. If over half the points in the batch are slightly outside the warning range, this may indicate a problem. Or if one point is well outside the warning

---

[12] USEPA, ORD, 1998. *Guidance for Data Quality Assessment: Practical Methods for Data Analysis.* EPA QA/G-9, EPA 600/R-96/084, January, 1998. Available on the Internet at www.epa.gov/ord/qa.

[13] USEPA, ORD. DQA DataQUEST, QA/G-9d, 1996.

[14] This interpretation is fostered by the term acceptance limit. A limit implies a finite endpoint. A better term is acceptance range, but this has not caught on in the environmental business.

range and the control range a problem may have been encountered with the particular sample. These are all signals that it is time to go look for reasons why the deviation from strict normality occurred.

Even when it is found that the quality control is definitely outside the control range for an identifiable reason, the associated analytical data may be reliable. An example is if the matrix spike on a sample gives a higher than normal recovery, but the sample result is below detection limit. Most reasons for the high recovery will not change the fact of the below detection limit result for the sample. Another example is when the quality control gives indication that the sample result is biased low, but the reported value for the sample is above the permit limit. Again, most reasons for the low bias are not going to eradicate the permit excursion. Other examples have been discussed in previous chapters.

The last point to be discussed about acceptance limits is the size of the range. As an example, you receive an accuracy result of 3% for bis(2-ethylhexyl)phthalate from a matrix spike on your effluent, and the acceptance range is presented as D-305 (detected to 305% recovery). A reasonable question to ask concerns the size of the range. Most users of laboratory data do not have the formal training or laboratory experience to look at a range and make the evaluation that it is too wide, or, if it is too narrow, for the target analyte. You might think that the method would be a guide to what is reasonable and what is not, but the information presented in the methods is frequently misleading or not relevant. Methods 624 and 625 present expected individual compound ranges for quality control check standards; however, the data were generated over 20 years ago for packed columns, and 99.9% of laboratories now use capillary columns. It can be reasonably expected that current ranges should be much tighter than those in these methods. On the other hand the acceptance range for matrix spike recovery (64-129) listed in the newest EPA method for oil & grease (Method 1664) was generated in research laboratories and is completely unattainable by the average analyst who processes a diversity of samples.

One approach to the problem is to go talk to the quality assurance officers in the lab and ask them to show you how the acceptance limits were generated. Look at the number of samples that are in the database used to generate the range. Ask if outlier detection and removal is used. Ask if the database contains only matrix spike results or if it also contains results from laboratory control samples. Ask how often the database is updated. Ask how far back does the database go? In brief, the answers should be that the database contains an absolute minimum of 20 matrix spike only data points that go back no more than 3 years. Outliers should be purged. The ranges should be updated at least annually.

Another approach is to compare the laboratory's range to a published range. As mentioned above, the Method is frequently not very helpful. The Air Force Center for Environmental Excellence (AFCEE) has lists of acceptance ranges that are generated from actual samples analyzed by many different laboratories. These are contained in the AFCEE Quality Assurance Project Plan, which can be found on the Internet at www.afcee.brooks.af.mil. Although these data are limited to methods found in the EPA Office of Solid Waste methods manual, SW-846, there is a lot of performance carry over to the similar wastewater methods. For instance, ranges generated from Method 8270, GC-MS analysis of semivolatile organic compounds, are reasonable guides to what can be achieved using the wastewater method 625.

CHAPTER 6

Instrument Quality Control

The quality controls that are performed in association with an analytical protocol can be divided into two groups. The first group are quality controls that are oriented toward evaluation of the sample and the entire analytical protocol. The second group of quality controls focuses upon the instrument that is used in the analysis. In previous chapters some of these instrument quality controls have been introduced. They include the performance of instrument detection level (critical level) studies, calibration, and calibration verification. There are other instrument quality controls that can be performed and examined. Retention time studies are designed to evaluate the variation of retention times. System performance evaluations examine the inertness, cleanliness, and response of the system. System performance also generates column condition information. The tune of mass spectral detectors can be evaluated.

It can be argued, with some justification, that the performance of calibration and daily calibration verification, are sufficient instrument quality controls. In fact there are many regulatory programs outside the environmental arena, where calibration and calibration verification are the only required instrument-oriented controls. Let's briefly examine calibration as an instrument quality control.

There are two types of calibration that are used. The first is based on the peak height while the second is based on peak area. The calibration by peak area tends to be the more robust of the two types. This means that the peak area calibration will still function even when there is considerable degradation of the liquid layer in the column, or active sites are present in the injector or column. Both of these problems result in the chromatographic peak spreading wider and developing an unsymmetrical shape (tailing). Loss of liquid layer will also lead to absolute retention times shifting to lower values, although relative retention times are generally unaffected. Column liquid layer degradation gives an increase in the baseline noise level. The degree of spread or tailing can even extend to the point where the peak exceeds the bounds of the retention time window. An instrument operator can manually adjust the beginning and the end of the peak to provide the correct area calculation. And of course, retention times are updated on a daily basis as part of the calibration verification process.

Another symptom of column degradation is decreased resolution. What were formerly resolved compounds develop overlap. Again adroit manipulation of the peak endpoints and baseline construction can give appropriate area counts for the overlapping analytes.

In both of these situations, the peak height of the analyte is affected in a non-correctable manner. Peak spreading means the height decreases while the width increases. Overlapping peaks can result in increasing peak heights. In either case no amount of legitimate manual integration manipulation will produce the needed peak height to give the "correct" quantitation. Thus calibration based on peak height does contain a degree of dependence upon system performance. Which is probably the primary reason that this type of calibration is not found very often in environmental laboratories.

The bottom line is that continuing calibration verifications based on peak area can be made to pass under situations where there is significant system impairment. Independent measures of system performance are essential information for the instrument operator as guides to maintenance decisions. System performance evaluations based on calibration have less usefulness.

A. System Performance and Maintenance

System evaluation can take many forms. The day-to-day tracking of analyte and internal standard retention times serves as an effective measure of column liquid layer loss. If two closely eluting analytes are tracked together, an evaluation of resolution ability is added.

Resolution can be evaluated as a specific control. The calculation of a resolution metric can be performed in a number of different ways. The most common formal calculation of resolution takes the form:

$$\text{Resolution} = \frac{2[RT_A - RT_B]}{W_A + W_B}$$

where W_A and W_B are the widths of the two peaks in minutes at the baseline and RT_A and RT_B are the retention time of the peaks in minutes. Values of resolution greater than 1.00 mean that the peaks are completely resolved at the baseline. Values less than 1.00 indicate overlap. The width of the peak at the half height point is sometimes a more appropriate and easier measurement to make, especially when the peaks exhibit some overlap. In this case the 2 in the numerator is deleted from the equation.

The analyst is not limited to this metric for resolution evaluation. An evaluation that takes into the account the depth of the valley between two overlapping peaks can be performed (Figure 6-1). A formula that embodies this idea is found in the Contract Laboratory Program Statement of Work for Pesticides. The depth of the valley from the peak apex of the second peak is measured, although a more appropriate evaluation would be from the apex of the smaller peak, since the second peak is not always the shorter.

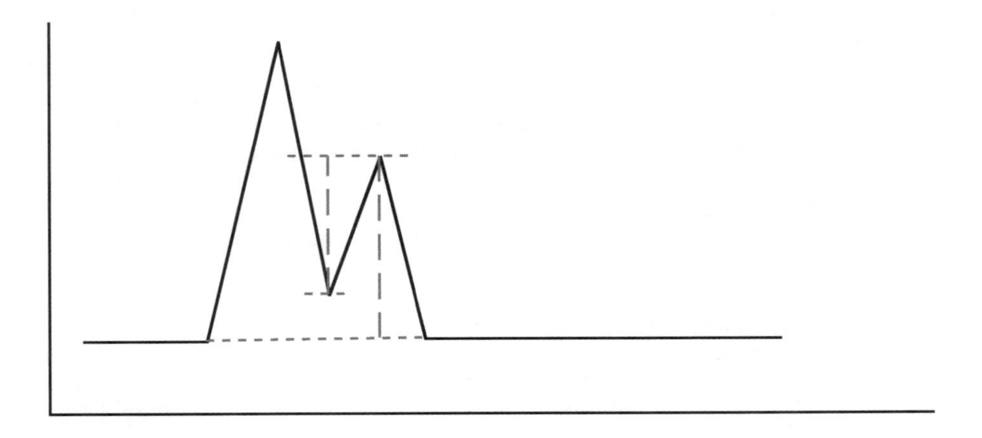

Figure 6-1. Measurements made to evaluate resolution by valley depth.

The valley depth is then divided by the height of the second peak to calculate percent resolution (% Resolution):

$$\% \text{ Resolution} = \frac{V}{H} \times 100$$

where V is the valley depth, and H is the second peak height from the baseline. In the CLP there are minimum values for this quality control, >60% for the resolution check mixture, a solution containing several pairs of pesticide target analytes that are expected to present resolution problems. As a quality control, the value of % Resolution will decrease as the column ages.

Some compounds will degrade along well-known characteristic pathways to measurable products. 4,4'-DDT and endrin are two such compounds. These compounds will degrade to give the products illustrated in Figures 6-2 and 6-3. The factors that contribute to the degree of DDT and endrin decomposition are the presence of active sites (dirt) in the injection port of the GC. The combination of heat in the injection port and acidic catalytic sites promote the chemical rearrangements. Systems that use direct on-column injection (rare in environmental work, but common elsewhere) are significantly less prone to breakdown even though the active sites may be present due to the much lower temperatures involved in the technique.

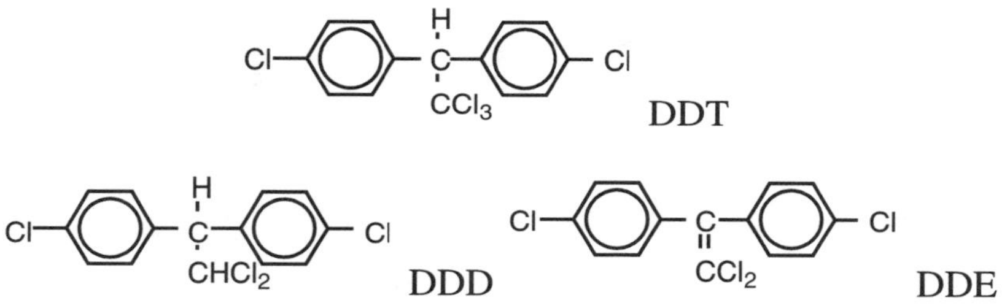

DDT

DDD

DDE

Figure 6-2. 4,4'-DDT and its two breakdown products.

Endrin

Aldehyde

Ketone

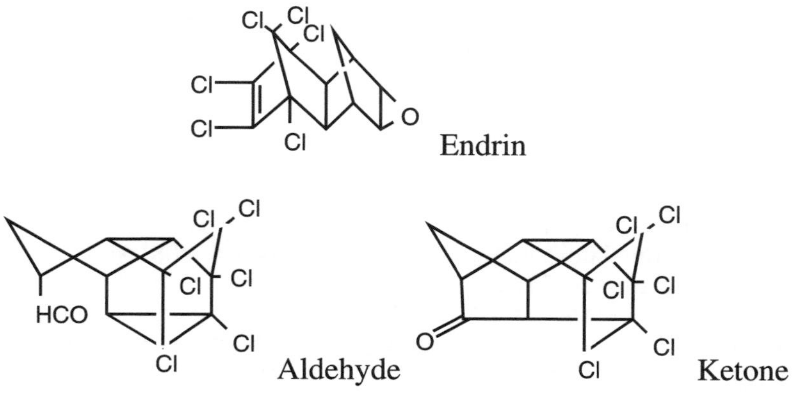

Figure 6-3. Endrin and its two breakdown products.

Breakdown is calculated for DDT and endrin by determining the amount of the parent and product compounds present from an injection of a solution containing only the parents:

$$\% \text{ DDT Breakdown} = \frac{\text{DDD} + \text{DDE}}{\text{DDD} + \text{DDE} + \text{DDT}} \times 100$$

$$\% \text{ Endrin Breakdown} = \frac{\text{Aldehyde + Ketone}}{\text{Endrin + Aldehyde + Ketone}} \times 100$$

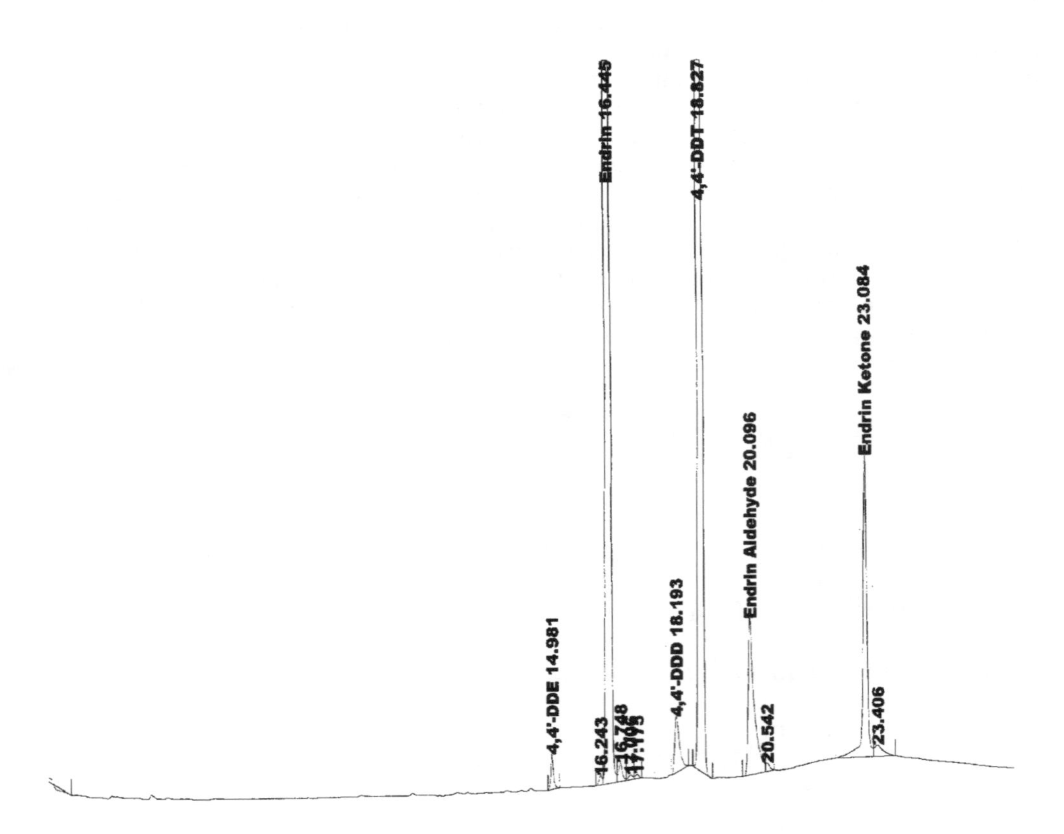

Figure 6-4. Chromatogram of a breakdown evaluation.

As the injector accumulates non-volatile residuals from sample injections, the percent breakdown will increase. Most regulatory methods (EPA Methods 608, 8081A, CLP-SOW) have a maximum cut-off for breakdown that ranges from 15% to 20%.

The maintenance to correct a breakdown problem is to replace the injector liner. Many laboratories will purchase and keep on-hand a supply of injector liners that are already cleaned and pre-silanized. This is a very expensive way to go about solving this maintenance problem. The glass liners for the injectors can be cleaned by brief immersion in an isopropanol-potassium hydroxide bath, followed by thorough rinsing with reagent water. This cleaning reagent neutralizes and removes the acid sites. After cleaning the surface of the glass must be made inert. Silanization reagents are used to cap the exposed -OH groups on the glass surface, which are a major cause of acidic active sites on the glass.

A number of silanization reagents are available, most useful as solutions in toluene. The silanization reagents that have chlorine attached to the silicon, such as dichlorodimethylsilane, are very quick-acting reagents, but the by-product of the reaction is hydrogen chloride, which can further react with the surface and must be rinsed off the liner thoroughly with solvent. Reagents that have no chlorine (*N, O*-bis[trimethylsilylacetamide], CAS 10416-59-8 or hexamethyldisilazane, CAS 999-97-3) are slower acting, requiring up to 30 minutes of contact time, but do not generate reactive by-products. The clean, oven-dried liner is immersed in the silanization reagent for an appropriate time, then removed and rinsed with clean solvent. The analyst must take care to avoid touching the liner with fingers or gloves - use only stainless steel forceps to manipulate the silanized liner. Storing the silanized liner can be a problem since it must be isolated from oxygen and moisture. The best practice is to keep the cleaned liners in a desiccator, then silanize and rinse them immediately before placing the liner in the injector port. The liner can be dried in the port under the flow of inert carrier gas if the column is not already attached. Otherwise it can be air dried briefly after rinsing, then put it into the injection port.

Glass injection port liners will last almost forever. By using the above simple procedure for cleaning and silanization, a stock of 30 liners can be continuously rotated through use and re-cleaning, allowing a fresh liner to be used every day of operation.

The calculation of tailing factors, i.e. an evaluation of peak shape, is a very specific indicator of active sites in the system. As illustrated in Figure 6-5, a line is constructed from the apex of the peak down to the baseline. A horizontal line is then constructed parallel to the baseline at a point equal to 10% of the peak height. The tail of the peak is then divided by the front of the peak to give the tailing factor. When there are chemically active (acidic or basic) sites in the system, compounds being chromatographed will bind to the sites. Increasing the column temperature will overcome the attraction of the analyte to the liquid layer; however, some of the molecules of the analyte will be bound to the active site. The active site-compound interaction is of greater strength than the liquid layer-compound attraction, and it takes more energy (heat) to overcome the active site attraction. Thus some of the analyte molecules will lag behind the bulk of the molecules in the peak. If the active sites present a very strong attraction for the analyte molecules, a distinct second analyte peak can be generated, with the size of the second peak being proportional to the number of active sites. This cause of peak splitting is generally exhibited by only the sensitive compounds in the analysis. When all the analytes in the chromatogram are exhibiting peak splitting this is commonly a symptom of an overloaded injection, a gap present in the liquid layer coverage inside the column, or a mechanical problem with the injector. In severe cases the active site-compound interaction can be so strong that the attraction is not overcome during the GC run and the analyte peak vanishes entirely.

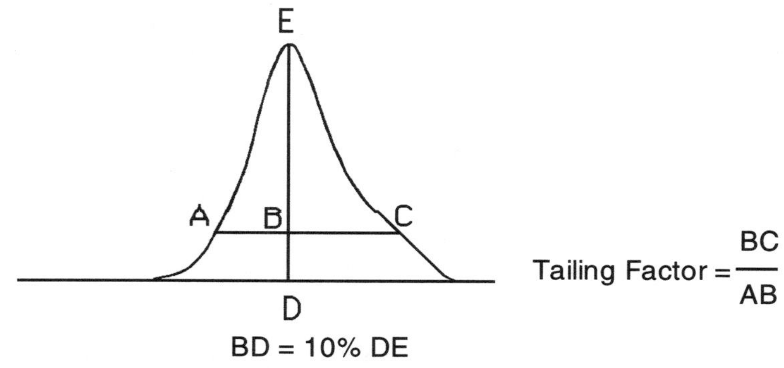

Figure 6-5. Tailing factor measurements and calculation.

The two compounds commonly used for tailing factor evaluations are benzidine and pentachlorophenol. Benzidine is an aromatic amino compound with base properties, while pentachlorophenol exhibits acidic qualities. These two compounds are more indicative of the build-up of active residues within the column than they are of the cleanliness of the injection port.

There are several maintenance steps that can be performed to remedy tailing problems. Assuming that the injection port liner is being changed daily, the first step is to remove the first foot or so of the column. Active sites are created by several routes. The first is due to the presence of a material in a sample that is transferred from the injection port to the head of the column, but, because of low volatility or strong attraction to the liquid layer, remains at the head of the column. This is more commonly seen when cool direct on-column injection techniques are being used, but it can occur with heated injectors particularly when crud in the injector is thermally decomposing. The second, and more important route is when a reactive material (water and/or acid) has been injected into the GC, and the liquid layer is decomposing to generate exposed silanol groups. In either case removing the first foot of the column physically eliminates the active sites.

The second step is to reverse the column in the GC. Take the injector end of the column and attach it to the detector and vice versa. When active sites are being created due to acid catalyzed decomposition of the liquid layer, the acid catalyst for this process will slowly progress down the column to affect the entire length of the liquid layer. Reversing the column allows the catalyst to exit the column to the detector rather than destroy the entire column.

The third step is to physically wash the column. The detector end of the column is attached to a reservoir of an organic solvent such as methylene chloride or toluene, then air pressure is applied to the reservoir to force solvent through the column. This allows organic materials to dissolve in the solvent and be removed. The liquid layer in the column consists of huge silicone molecules with organic appendages, and they are not soluble to any significant degree in the solvent.

B. Tuning

The tuning of the mass spectrometer is essential for its use as an analytical tool. The key elements in the operation of the mass spectrometer are the calibration of the mass axis and achieving unit mass resolution. The calibration of the mass axis means that ions of a certain m/z are actually detected at that m/z. Unit mass resolution means that the instrument can distinguish between adjacent m/z signals, for instance 441 and 442. If the mass axis is calibrated correctly and unit mass resolution is achieved, most mass spectrometers will generate mass spectra that are reliably usable for at least qualitative analysis. Other factors, such as the relative abundance balance between the low and high ranges of the m/z scale, help to achieve quantitatively reproducible results, but they are window dressing compared to the key factors.

There are two types of tune that are performed. The first is to a material, perfluorotributylamine (FC-43, PFTBA, CAS 311-89-7) that has become a de facto standard among the instrument suppliers. Most manufacturers have a small vial of PFTBA attached to the side of the mass spectrometer source. A solenoid valve allows bleeding small amounts of this material into the source so that the tune can be performed. The operator (or more commonly an automated tuning program) adjusts the various electrical and magnetic field strengths and cycling rates in the instrument to achieve an appropriately appearing spectrum. This tune is performed normally at least once each day of instrument operation.

The second type of tune is a characteristic of the USEPA Methods for analysis of compliance monitoring samples. Specific tuning compounds are injected into the

GC-MS every 12 to 24 hours of instrument operation, decafluorotriphenylphosphine (DFTPP, CAS 5074-71-5) for semivolatile analysis and bromo-4-fluorobenzene (BFB, CAS 460-00-4) for volatile analysis. The spectrum of the tuning compound is then evaluated qualitatively and quantitatively against method-specific acceptance ranges. The mass axis calibration, unit mass resolution, and overall spectral mass balance are evaluated, similar and additional to the evaluation given to PFTBA. However, the influence of the gas chromatograph upon the peak shape and results is also indirectly evaluated by these tuning procedures.

Many games have been played by both laboratories and data reviewers over the use of "correct," i.e. method specified, tune acceptance criteria. Particularly obvious to operators that had to manually tune their instruments, after you optimized the instrument for response to PFTBA you had to "detune" the MS to meet the requirements of DFTPP or BFB. Many older operators still remember their particular set of tricks for distorting the PFTBA spectrum so DFTPP would pass. The modern versions of the instrument software automatically adjust the MS parameters for PFTBA to meet DFTPP requirements. In the long run, assuming that these additional tuning checks actually serve some purpose, which is doubtful, it does not matter from a scientific viewpoint which set of tune criteria are used. As long as the mass axis is calibrated, unit resolution is achieved, and if there is a modicum of low-to-high mass balance, the instrument will work. Method compliance issues are primarily the driving force for the continued use of DFTPP and BFB tunes as instrument performance measures.

CHAPTER 7

Laboratory Contamination

All organic analysis laboratories have contamination. There is contamination originating from the solvents, reagents, and standards. There is contamination introduced by the consumable supplies and equipment. A major source of contamination is the samples that are analyzed.

Sometimes the contamination is present in the form of the target analytes. More frequently the contamination is not a target analyte; however, it contributes to the background noise and miscellaneous unidentified peaks that are present in the chromatograms.

There is contamination (systemic) that is found at approximately the same level in every single analysis. And there is mysterious contamination (random) that expresses itself only occasionally.

The bottom line is that all laboratories have contamination problems. One task of the data reviewer is to determine how diligently the laboratory is evaluating, tracking, and minimizing contamination and then to apply this information to decisions about which reported substances in the sample are present in the field source and which should be attributed to laboratory contamination.

There are a number of papers in the literature where erroneous conclusions are drawn based on results derived from sloppy lab work and rampant laboratory contamination. Sometimes these papers are rebutted[1], but more commonly they remain unchallenged. Further the rebuttal is frequently printed in another issue of the journal, and without knowledge of the existence of the rebuttal, future researchers will cite the original paper as a justification of some special property of the laboratory contaminants. (As an editorial comment: there is nothing exceptional about methylene chloride and bis(2-ethylhexyl)phthalate, other than they are the two most frequently encountered laboratory contaminants. They degrade and can be remediated just like other similar organic compounds.)

The most common quality control tool that is used for assessing laboratory contamination is the blank. A blank is performed by taking an aliquot of analyte-free water or solid through the entire sample preparation and analysis protocol, then examining the chromatogram and quantitation report for the presence of contamination.

Almost all laboratories perform blanks because they are part of the required method protocols. Unfortunately, the data from the blank are frequently recorded, filed, and forgotten. Very few laboratories use the data that are present to the fullest extent possible as aids to identify, evaluate, and track contamination.

The reported results on an individual blank can lead to a false sense of security. The treatment given to a blank may be very different than that accorded the sample. Blank and sample results can be manipulated to make it appear that no contamination problems exist when, in fact, they are rampant.

There are alternate data sources in analytical batches that allow evaluation of contamination levels. Examination of duplicate, matrix spike, laboratory control, and continuing calibration results can give clues about laboratory contamination.

---

[1] Smith, R.-K., 1997. Discussion of: Organic Priority Pollutants in New York City Municipal Wastewaters: 1989-1993, A.I. Stubin, T.M. Brosnan, K.D. Porter, L. Jimenez and H. Lochan, *Water Environ. Res.* 1997 69(3):382 May/June 1997 issue.

Comparison of sample results from different analytical batches can lead to identification of contamination. And sometimes the principle of Ockham's Razor (the most probable answer) is the necessary insight that leads to the determination that contamination is present.

There is a standard list of the usual suspects that are frequently present as laboratory contamination. The list includes methylene chloride, toluene, xylenes, acrylates, phthalates, naphthalene, acetone, MEK, MIBK, diacetone alcohol, BHT, cyclosiloxanes, and several others (Table 7-1). But any compound can be present as laboratory contamination. This chapter is intended to lead one through the maze of laboratory contamination.

Table 7-1. The list of usual suspects found as laboratory contamination.

Usual Suspect	CAS #	Contaminant Source
Phthalates		
diethyl	84-66-2	Plasticizer, perfume solvent
dibutyl	84-74-2	Plasticizer, perfume solvent
benzylbutyl	85-68-7	Plasticizer
bis(2-ethylhexyl)	117-81-7	Plasticizer
dioctyl	117-84-0	Plasticizer
Methylene chloride	75-09-2	Laboratory solvent
Toluene	108-88-3	Laboratory solvent, petroleum fuels
Xylenes	106-42-3	Deionization resin breakdown, petroleum fuels
Acrylates		Fingernail polish and carpet glue monomers
methyl	96-33-3	
ethyl	140-88-5	
Methacrylates		Fingernail polish and carpet glue monomers
methyl	80-62-6	
ethyl	97-63-2	
Styrene	100-42-5	Deionization resin breakdown
Acetone	67-64-1	Laboratory solvent
Naphthalene	91-20-3	Toilet disinfectant and moth balls
Methylethyl ketone (MEK)	78-93-3	Laboratory solvent
Methyl isobutyl ketone (MIBK)	108-10-1	Laboratory solvent
Diacetone alcohol	123-42-2	Aldol condensation product of acetone
5,5-Dimethyl-2(5H)-furanone	20019-64-1	Aldol condensation product of acetone
Butylated hydroxy toluene (BHT)	128-37-0	Food preservative
Cyclosiloxanes		GC column liquid layer breakdown
hexamethylcyclotrisiloxane	541-05-9	
octamethylcyclotetrasiloxane	556-67-2	
Hexane	110-54-3	Laboratory solvent
Aliphatic hydrocarbons	-	Petroleum fuels

Table 7-1. The list of usual suspects found as laboratory contamination, *continued*

Usual Suspect	CAS #	Contaminant Source
Cyclohexene and oxidation products	110-83-8	Preservative in methylene chloride
cyclohexanone	108-94-1	
cyclohexeneone	930-68-7	
cyclohexenol	822-67-3	
chlorocyclohexene	930-66-5	
chlorocyclohexanol	1561-86-0	

A. Blanks

A blank is an analyte-free portion of reagent water or solid (sodium sulfate or sand) that is manipulated exactly like a sample through the preparation and analysis protocol. All steps of the protocol must be included. For batches that are not processed during a single work shift (open or add-on batches) additional blanks must be included each time samples are prepared.

If a clean-up step is included for some of the samples in the batch, then a separate clean-up blank might be used to identify the possible contamination arising from just the clean-up step. Gel permeation chromatography (GPC) is a clean-up where the analyst must use a separate blank. Bis(2-ethylhexyl)phthalate, methoxychlor, and perylene are target analyte compounds in the calibration mixture for GPC and there is tremendous potential for sample contamination with these materials. Both the batch blank and a separate GPC blank should be taken through the GPC as additional assurance that the samples are not being infected. The GPC blank can be processed immediately after the completion of the calibration, then the batch blank should be processed at a random position later in the GPC run.

This idea can also be applied when samples are filtered to remove particulates using a syringe filter, or subjected to absorption chromatography on disposable columns. Over the last several years manufacturers have made great improvements toward providing contamination-free disposable media, but there are still items on the market that are unsuitable for trace analysis. Many items, particularly syringe filters, are designed for and marketed to clinical laboratories, and definitions of suitability and cleanliness are different for those applications. A device may be sterile, but completely inappropriate for chemical laboratory use. One specific problem is the widespread use of bis(2-ethylhexyl)phthalate as an additive in a majority of flexible plastics. This plasticizer is considered completely safe and appropriate for medical uses, such as in flexible blood bags and tubing (Tygon®), but is the bane of trace analysis laboratories. A blank can be filtered through a randomly chosen filter from a newly received lot number to check the cleanliness of the supplier's lot of the devices prior to use on samples, then the batch blank used as a continuing check.

The single blank constitutes a spot check for laboratory contamination at a specific point in time for a specific set of laboratory operations. It is useful as an evaluation of systemic contamination, but rather restricted in providing definitive data about overall contamination levels.

To get the most use out of blanks, they must be evaluated in sets using statistical techniques. Reliable statistical evaluations are possible only when all the data are present. Many instrument operators have a habit of setting the report generation software to report only those target analytes that are present at or above the method detection level or the quantitation level. For example, if the quantitation level for target analytes is 10 ug/L, the software can be set to report only those analytes at or above 10 ug/L. The analyte is deleted from the printed report if the quantitation for the analyte is below 10 ug/L. But this censoring can lead to misrepresentation of the data and an obvious skewing of the statistical evaluation.

The misrepresentation comes about when there are target analytes present in the blank at levels just below the quantitation limit, but above the limit in samples. For instance, suppose the blank has bis(2-ethylhexyl)phthalate at 9 ug/L (by default listed and treated as zero on the quantitation report), the sample has 11 ug/L, and the quantitation limit is 10 ug/L. Reporting a hit in the sample for bis(2-ethylhexyl)phthalate at 11 ug/L without qualifying the result as probably due to blank contamination (B) is misrepresentation of the data. This may not be an intentional misrepresentation on the part of the instrument operator, because the software will omit listing the compound on the quantitation report.

This practice also produces problems in statistical evaluations. Contaminant compounds above the quantitation level are entered into the statistical database as a numerical amount. But what is to be done with the blanks where the contaminant is not listed? It is impossible to enter a "BDL." Instead a number must be entered. Zero is a very, very bad choice as a replacement for a "BDL." Many statisticians will enter a value that is half of the quantitation limit, 5 ug/L in this case. But this is also a bad choice. Assuming that laboratory contamination is somewhat under control, the majority of the data points will have a value of 5, with occasional points at or above 10. The statistics are skewed and worthless.

The solution is that quantitation reports of blanks must be generated without any data censoring. If the software detects a signal, regardless of size, in the appropriate retention time window, a quantitation must be allowed and accepted. For the normal target analyte list this means that the quantitation report may go on for several pages with a lot of low level quantitations that are nonsense. But these values are essential for a valid statistical evaluation.

The evaluation commonly used is the control chart. Figure 7-1 presents a control chart of bis(2-ethylhexyl)phthalate from several months of blanks from an EPA Method 8270C procedure. Note that the analyst used a 1 ng/uL report cut-off up until September, 1999. The 15 lowest points up to this time are assigned the value of 0.5 ng/uL. After this time, the software settings were changed to allow reporting lower values. Although this practice did not significantly change the chart for bis(2-ethylhexyl)phthalate, in the following control chart for dibutylphthalate (Figure 7-2) it significantly influenced the data.

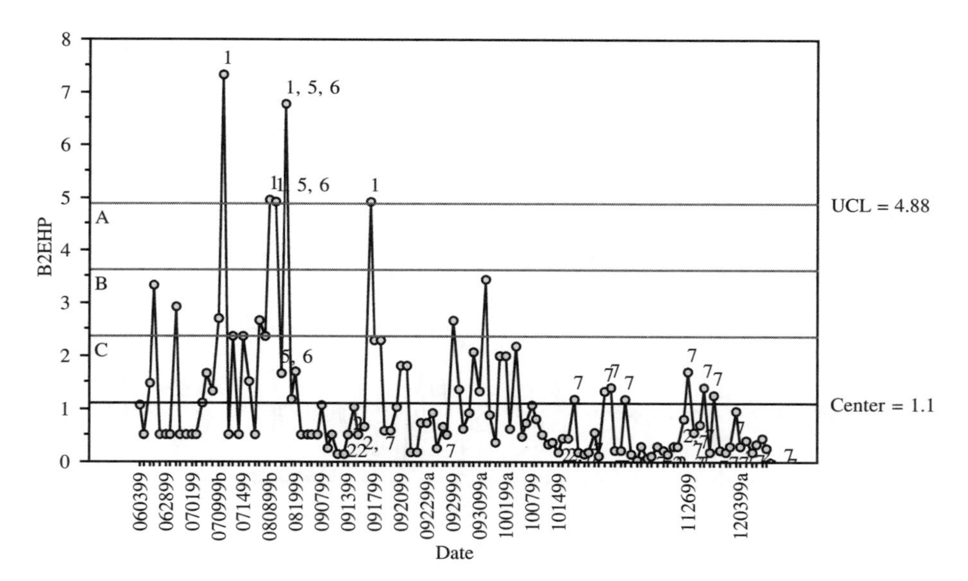

Figure 7-1. Control chart of bis(2-ethylhexyl)phthalate contamination levels. Units are ng/uL injected on-column.

There are a number of advantages that accrue from this type of data treatment. The first and foremost is that the traditional rule-of-thumb requirement for "sample data must be greater than 10 times the amount found in the blank, otherwise blank contamination is assumed," can be replaced with a statistically verifiable data analysis. In this example the control limit ($\pm$3 sd) for bis(2-ethylhexyl)phthalate is 4.88 ng/uL, no points above 8 ng/uL are present, and this suggests that a reportable quantitation level of 10 ng/uL for samples can be used without any blank qualifiers.

Second, the presence of bis(2-ethylhexyl)phthalate contamination tends to be significant only during limited periods of time. Efforts to eliminate and minimize contamination can be focused upon what was different during these times. In this case the time periods were found to be associated with the use of a particular substitute technician, while one of the regular technicians was out. The control chart illustrates that as the substitute technician obtained more practice, his technique, reflected in the decreasing levels of contamination, improved significantly.

The third advantage is that quantitative feedback can be provided to the sample preparation technicians, instead of the normal, "You botched it. Go re-extract that entire batch." These data can be used for training and competence evaluations.

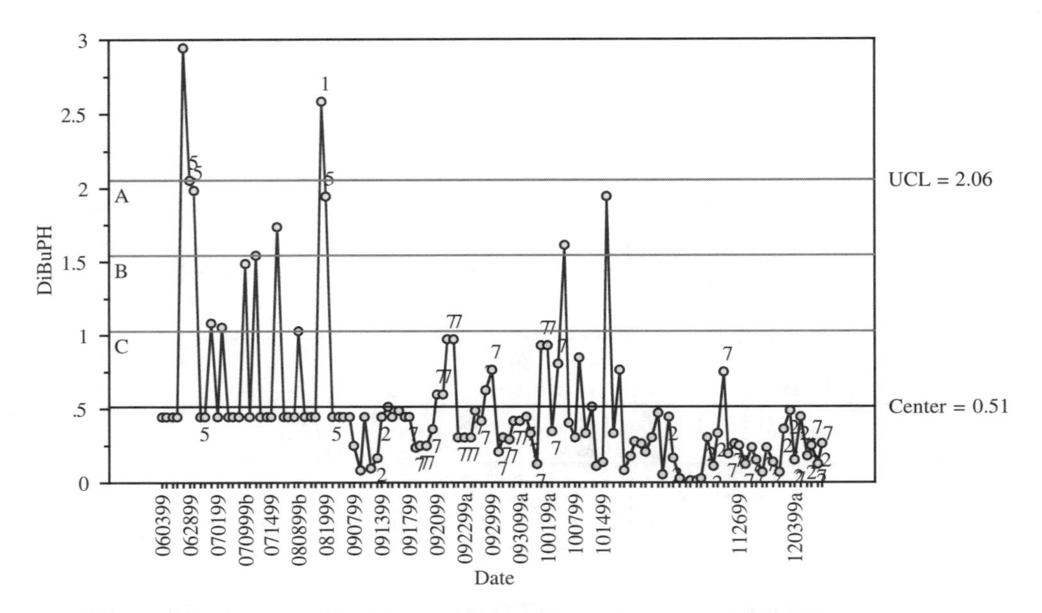

Figure 7-2. Control chart for dibutylphthalate in blanks. Units are ng/uL on-column.

B. Phthalate Contamination

Phthalates are a continual problem in the semivolatiles laboratory, and some management efforts need to be focused upon reducing the presence of these compounds. The first area is to identify and eliminate or minimize the use of materials that contain phthalates. Flexible polyvinyl tubing (Tygon® brand), which contains 40% by weight bis(2-ethylhexyl)phthalate, is the number one suspect. Even the briefest contact of an organic solvent such as hexane or methylene chloride will extract milligram amounts of bis(2-ethylhexyl)phthalate from this tubing. Manipulating the tubing with a pair of gloves can transfer several milligrams to the gloves. Use of this tubing to connect a nitrogen cylinder to a blow-down needle will guarantee contamination levels above 10 ng/uL in the final sample extract.

Tubing is not the only source of phthalate contamination. The plastic bags used to containerize bulk amounts of sodium sulfate drying agent are loaded with phthalates. Plastic shrink-wrap around disposable pipets and other items contributes phthalates to the overall contamination levels. Most medical examination gloves are coated with phthalate-containing lotions for the comfort of the user. Scented paper towels, dish washing formulations, and hand lotions use phthalates for the perfume carrier. The list goes on and on.

The most reliable check that can be made is to rinse or soak the item in methylene chloride, then assay the concentrated rinseate for phthalates.

For some reason that is entirely unfathomable to laboratory technicians, there seems to be an inverse relationship between phthalate contents of consumable supplies and cost. A technician can make the effort to check for the presence of phthalates in gloves, for instance, and find a brand that is suitable. But these turn out to be the most expensive available, so, unless the laboratory personnel explain to the purchasing manager the basis for their recommendation, the purchasing manager may buy the lowest cost gloves, which invariably are loaded with phthalates. Laboratory personnel always have to explain their recommendations to purchasing personnel, otherwise purchasing personnel make decisions solely on the basis of cost.

Elimination of all sources of phthalate contamination is not possible, so a back-up measure must be used to insure minimization of the problem. A rigorous glassware cleaning procedure helps. In most laboratories the first step is to remove organic residues from the items with a solvent rinse and adhering materials with a brush and water. A very effective means of removing trace organic residues is to soak the solvent-rinsed glassware in a solution of isopropanol and potassium hydroxide. Potassium hydroxide will dissolve in 100% isopropanol to the extent of about 20% by weight. Using the commonly available 70% isopropanol-water is not as effective, but up to about 50% by weight potassium hydroxide will dissolve in it. This solution is very corrosive, even to the extent of dissolving glass, so glassware can not be left in it for any prolonged periods. A one-hour soak will remove and destroy almost any organic residues. Phthalate esters are quickly and completely hydrolyzed in this solution. Volumetric flasks should be only briefly (about 5 minutes) submerged. Volumetric pipets should not be treated with this solution, rather they should be submerged in a strongly acidic solution of a cleaner such as No-Chromix®. The isopropanol-potassium hydroxide soak can be prepared and used in a high density polyethylene or Teflon bucket. Depending on the rate of use, the solution will last about one week. After soaking in the solution the item is thoroughly rinsed with reagent water.

Next the item is soaked in a heated water solution of a detergent. It is important to find a detergent that does not contain phthalates in the formulation or packaging. If the laboratory is plumbed for steam, a heated soak can be created from a large stainless steel, glass, or Teflon bucket (at least 5-gallon capacity or larger) that contains a generous coil of copper tubing that is attached to the steam outlet. The water and detergent are mixed in the bucket, the glassware added, then the steam turned-on to raise the temperature up to almost boiling (an immersion heater can be used as an alternative). The glassware (including volumetric glassware) is allowed to soak for several hours, then removed while hot and thoroughly rinsed with reagent water. This detergent soak should be made fresh each day.

Almost all laboratory glassware will be squeaky clean after isopropanol-hydroxide and heated detergent soaks. However, just to be sure, immediately prior to use, the glassware is rinsed with the solvent that is going to be employed in the sample extraction.

C. Methylene Chloride Contamination

The most significant source of methylene chloride (dichloromethane, CAS 75-09-2) contamination in volatiles analysis is the semivolatiles laboratory where it is the most widely used extraction solvent. More than any other factor, the solution to methylene chloride contamination is an engineering and laboratory layout challenge. Methylene chloride is a very volatile compound, and as it is used it quickly builds up significant concentrations in the air. Highly efficient air venting systems can minimize the air concentration in the semivolatiles laboratory, but will never completely eliminate methylene chloride. The only way to minimize methylene chloride in the volatiles laboratory is to first physically separate the two labs, second, provide separate HVAC systems for the two labs, keeping the volatiles lab under positive air pressure, and, finally, prohibit people traffic between the two labs. Methylene chloride will absorb to a person's clothing in the semivolatiles lab and then desorb in the volatiles lab.

Unlike most other laboratory contaminants, methylene chloride tends to act like a systemic contaminant. The amount found in any contaminated sample or quality control is related to the length of time the sample is exposed to the air in the laboratory. If the air is heavily contaminated, the result will be a constant presence of relatively high levels (>50 ug/L) methylene chloride in every sample. The amount may be observed to cycle in time with the solvent activity in the extractions lab. In laboratories that practice contaminant control, the amount of methylene chloride decreases overall, and the compound begins to behave more as random contamination. In either case control charts can be used very effectively to monitor the methylene chloride levels and provide guidance for sample result evaluations.

D. Acetone Complex

Acetone, isopropanol, diacetone alcohol, 5,5-dimethyl-2(5H)-furanone, and 4-methyl-3-pentene-2-one are related to each other and are found as contaminants in both the volatiles and semivolatiles laboratory. Isopropanol (isopropyl alcohol, 2-propanol, CAS 67-63-0) and acetone (2-propanone, CAS 67-64-1) are related by oxidation/reduction and any sample of one is contaminated with varying amounts of the other. Diacetone alcohol (4-methyl-4-hydroxy-2-pentanone, CAS 123-42-2) is the aldol condensation product of two molecules of acetone and is present to one degree or another in all samples of acetone. 4-Methyl-3-pentene-2-one (mesityl oxide, CAS 141-79-7) is produced from diacetone alcohol by the loss of a molecule of water. 5,5-Dimethyl-2(5H)-furanone (CAS 20019-64-1) is the cyclization product of diacetone alcohol. Acetone is a target analyte in the volatiles lab, while the other three compounds can be found in both volatiles and semivolatiles analyses as tentatively identified compounds (TIC). 4-Methyl-3-pentene-2-one and 5,5-dimethyl-2(5H)-furanone are volatile and can contribute to airborne contamination.

Acetone is widely used in the organics laboratory as a rinse solvent in glassware cleaning. Being miscible with water, this serves to quickly dry the glassware, and because it is a powerful, polar solvent it removes traces of detergent and any organic materials. The acetone is allowed to evaporate from the glassware, which does two things. First, it increases the acetone levels in the air, where it behaves much like methylene chloride as a volatiles contaminant. Second, diacetone alcohol is much less volatile than acetone, so any diacetone alcohol in the acetone is left as a concentrated film on the glassware.

Isopropanol is frequently used as a final rinse in the field decontamination of sampling equipment. Old bottles of isopropanol will frequently contain substantial amounts of acetone and diacetone alcohol. Use of isopropanol will commonly leave films of diacetone alcohol on the equipment that can be detected in the equipment blanks.

When acetone has been used in the semivolatiles laboratory, virtually every sample that is analyzed will have detectable amounts of diacetone alcohol, and it shows up in the TIC reports. In almost all cases the data reviewer should discount the presence of diacetone alcohol as representative of the pollutants at the field source. Confirmation of diacetone alcohol on-site requires elimination of the use of acetone and isopropanol in all aspects of the analytical procedure from sample container and glassware cleaning, to instrument calibration and sample analysis.

E. Blanks in the Volatiles Laboratory

The results from a blank can be somewhat misleading in the determination of contamination in the volatiles laboratory. One of the problems is that the regular batch blank is handled differently than most samples. The major differences lie in the source of the water or solid that is used to prepare the blank and how they are stored.

Tap water that is suitable for drinking is completely unsuited for any use in the volatiles laboratory unless it has been further purified. Analysis of tap water generally reveals the presence of the trihalomethane disinfection by-products from the chlorination process: chloroform, dichlorobromomethane, dibromochloromethane, and bromoform. If any iodine is present in the source water, trihalomethanes with iodine substitution will also be found. The drinking water regulations allow up to 100 ug/L total trihalomethanes, and in systems that are using chlorination, totals of up to 80 80 ug/L are frequently found. These amounts are well above the permissible levels for reagent water, where values of 2 ug/L are indicative of significant contamination. Other materials such as toluene are frequently encountered in drinking water. Bottled water is not an acceptable substitute since most bottled water is simply tap water placed in jugs. The laboratory must prepare its own reagent water.

Reagent water that is used to prepare the blank is commonly obtained from a commercial water purification system that consists of deionization and carbon filtration cartridges that may be coupled with UV treatment, distillation, or ultrafiltration steps. Deionization is necessary because metal cations interact with many organic compounds to form stable ligand complexes that are not dissociated under the conditions of the purge & trap or headspace sample preparation. As they age, the deionization resins decompose, frequently through a de-polymerization process, releasing organic compounds into the reagent water. The most commonly encountered resin decomposition compounds are *p*-xylene, styrene, and 1,4-divinyl benzene, although others on occasion can be present.

The initially prepared reagent water is heated to boiling, then purged with purified nitrogen or helium while cooling. Purging by itself is not sufficient, particularly when the resins in the deionization system are decomposing. Boiling is necessary to drive out the heavier volatiles. The purging serves to protect the water as it is cooling from re-absorbing atmospheric contamination. The purge should be vigorous. A slow purge with large bubbles is next to useless. Instead a fritted purge tube should be used to generate a large number of very small bubbles, and the volume of gas passed through the water should be sufficient to fill the container with bubbles. A flow of at least 1 L/min is needed. Reagent water for blanks and sample dilutions must be prepared fresh every day.

When the batch blank has hits for the resin decomposition products, it is rare for the associated samples in the batch to exhibit the same hits unless the samples have been diluted with the reagent water. Even for soil samples that have been prepared and preserved in acidified reagent water (EPA Method 5035), there is generally a time lag between preservation of the soil and the associated analysis, so that the reagent water analyzed today as the blank is not the same reagent water that was used to prepare the preservation solution. In this case it is necessary to examine the reagent water blank from the day the samples were preserved.

Samples that contain high concentrations of volatiles are a significant source of contamination. A container such as a 4 oz. wide-mouth jar is never hermetically sealed and will continuously leak volatiles into the surroundings. Refrigerators used to store samples concentrate the leaking compounds, which can permeate even septum-sealed containers. The solution is two fold. First, never store potentially high-level samples with low-level "clean" samples. Samples in jars cannot be stored with VOA-vial or EnCore® containerized samples. Have three or more sample storage refrigerators for the different types of samples: low-level water, low-level soils, and potentially high-level samples. Second, monitor the contamination levels in the storage refrigerators on a regular basis. Place a VOA-vial filled with reagent water in each refrigerator. On a weekly basis change and analyze the refrigerator blank for target and non-target analytes. When hits are found in the refrigerator blank, check all the samples that have been in the refrigerator during the preceding week (and those that may still be there) for similar hits. If any are found the results should be flagged with a "B" for possible blank contamination. Compounds that are most frequently found in refrigerator blanks are those associated with petroleum hydrocarbon fuels such as benzene, toluene, xylenes, other small aromatic hydrocarbons, and aliphatic hydrocarbons. Generally more than one compound will be found. This is not to say that no other compounds will ever be found, as any volatile substance in relatively high concentrations in a sample can contaminate other samples.

F. Laboratory Contaminants That Aren't

As a general rule, hits for phthalates, methylene chloride, acetone, and the other usual suspects reported from laboratories send up red flags for the data reviewer. These materials are not all that frequently encountered in field sources, particularly when they are the only contaminants found and reported. There are other compounds that are occasionally reported by laboratories that are, as a rule, completely unexpected from a site. Confronted with these particular hits, the engineering firm, being at a loss to explain them, blames the laboratory for laboratory contamination. Chief among these compounds is carbon disulfide (CS_2, CAS 75-15-0).

Carbon disulfide is a volatile liquid and is determined by purge & trap GC-MS. It is occasionally used in some industries as a solvent, and in others as a reagent, but overall its use is quite restricted. EPA has listed carbon disulfide as a hazardous material, and it is included as a target analyte on the Appendix IX groundwater monitoring list.

In the environmental laboratory, carbon disulfide finds few uses. The chief use was as the extraction solvent in the old California diesel range organics (DRO) method; however, this has been replaced in most labs by procedures that use methylene chloride for extractions. There are a few NIOSH methods that still require carbon disulfide for leaching target analytes from adsorption tubes. Both of these types of procedures are performed in the semivolatiles laboratory, while carbon disulfide is a target analyte in the volatiles laboratory. If there are normal laboratory contamination controls in place for volatiles, there should be no problem with carbon disulfide.

Where the problem arises is that carbon disulfide is a naturally occurring compound. Sulfur is an essential element for life, and is found in many proteins and other biochemicals. Similar to the production of carbon dioxide and water in oxygen-based environments, respiration in sulfur-based environments produces carbon disulfide and hydrogen sulfide. They are the metabolic end-products in anaerobic ecologies where sulfate is being used by microorganisms as the major electron acceptor in respiration. Another characteristic product of sulfur-based ecologies is methane. Examples of the microorganisms utilizing these metabolic processes include those of the genus *Desulfovibrio* and *Desulfuromonas*. These type ecologies are physically characterized as low-lying, stagnant, wet areas. Both water and sub-surface soil samples taken from these areas will frequently exhibit detectable amounts of carbon disulfide. There are a number of other sulfur-containing materials also found in these samples, such as thio-ethers, mercaptans, disulfides, trisulfides, and elemental sulfur, but none of these are target analytes of the EPA Methods used for site evaluations. Analyzing the sample extract with a sulfur-sensitive GC detector such as the flame photometric detector (FPD) in the sulfur mode will normally reveal numerous sulfur-containing peaks. Unfortunately most environmental laboratories do not include an FPD in their equipment inventory.

Engineering firm project managers, confronted with reported hits of carbon disulfide in samples and no historical record of its use on the site, frequently lack any background in microbiology and thus are unaware of the natural occurrence of this compound. The common solution seems to be to resort to the tried and true technique of blaming the problem on the laboratory, even when the laboratory produces verifiable evidence that there is no in-house contamination due to carbon disulfide.

Another example where laboratory contamination was a convenient excuse occurred during a project where explosives residues were being monitored in soil and groundwater. The regulatory agency was requiring detection limits for nitro-aromatic compounds that were lower than the capability of the HPLC being used. The chromatograms of the samples were exhibiting considerable noise on the baseline, and it often happened that random noise generated signals on both the analytical column and confirmation column within the proper retention time windows for some of the target analytes. The laboratory was forced by the regulators and the engineering firm to report these spurious signals as hits. Then in the final report the engineering firm turned around and blamed the laboratory for contamination problems when the "hits" did not match-up with site maps of groundwater flow patterns and known contaminant plumes.

There are other compounds that occasionally will be explained away by claiming laboratory contamination. The ability of the laboratory to defend itself from these charges frequently depends on the existence of a vigorous and documented laboratory contamination monitoring program, and having persons on staff that are knowledgeable of many other facets of science in addition to chemistry.

G. Other Contaminant Indicators

Blanks are the most widely used tool for detection of laboratory contamination, but they are not the only available tools. Anytime that a sample or quality control has been analyzed in duplicate, an opportunity presents itself for contaminant monitoring. These opportunities must be exploited.

The best alternate check point is the laboratory control sample. These are particularly information rich when an abbreviated list of analytes is being used, such is suggested in the EPA 8000-series (SW-846) methods. The only compounds that are supposed to be in these samples are the surrogates, internal standards, and a few target analytes. Additional hits in the quantitation report besides these compounds are generally due to laboratory contamination. Visually the chromatogram should display a completely flat baseline, marred only by the intentionally added compounds. Any little or big blips on the baseline suggest the presence of laboratory contamination. These can be tentatively identified and labeled as contaminants. If the laboratory control samples have been analyzed in duplicate, then comparison of the two samples from each batch gives an indication of the presence of random contamination. Otherwise, examination of lab control samples over many batches can give an indication of the efficacy of contamination control. The laboratory control sample is a powerful tool for monitoring laboratory contamination since it generally is treated just like every other sample, whereas the blank may be given somewhat special treatment.

Next on the list of alternate checks are the field blanks and equipment rinsates. These are not as definitive as the laboratory blanks or the laboratory control samples, because there are influences from the handling of these samples in the field in addition to the in-laboratory manipulation. What can be found in these samples is either a confirmation of the presence of contamination arising in the laboratory or an indication that contamination is being introduced in the sampling process. However, contamination present in the field blanks or equipment rinsates without confirmation from the lab control samples or lab blanks is fairly good evidence that there are problems with the sample collection protocol.

Sample duplicates and matrix spiked samples that are analyzed can give an indication of the presence of random laboratory contamination. These checkpoints are not as sensitive as the regular quality controls, due to the presence of increased chromatographic background levels; however, evidence for significant contamination problems can be exhibited in these samples. Data derived from these sources must be interpreted with a large measure of caution. One of the common interpretational failings of data reviewers is when they treat reported quantities of target analytes in samples and duplicates as absolute verifiable numbers without taking into account normal analytical result variation. For example, if a sample has a hit for methyl ethyl badstuff at 14 ug/L, and the sample duplicate is reported as <10 ug/L, there is no justifiable reason for raising a stink about laboratory incompetence. Experienced reviewers append a result variation of ± the quantitation limit to the results before they voice such worries. In this case the comparison is between 14 ± 10, and <10 ± 10, and considerable overlap is seen between these ranges. However, if the values were 35 ± 10 and <10 ± 10 then there may be cause for some concern.

Sample comparisons can also be performed when the samples are collected in approximately the same area in the field and analyzed either together in the same batch or on different days in different batches. As described above, these comparisons must be interpreted with regard for analytical variability, but they can still be very useful as indicators of significant laboratory contamination problems.

Repeating what was said at the beginning of this chapter, all organic analysis laboratories have contamination problems. The mark of the good laboratory that produces reliable data is the presence of a verifiable program to monitor and control laboratory contamination.

CHAPTER 8

Assessing Target Analyte Recovery

I love the idea of this chapter, and have been wanting to write it for the last eight years, which covers four different books (one in the fourth edition), six laboratory manuals, and 17 sessions of my 16-hour data evaluation course. It's the response to the two ultimate questions in analysis: "Is the target analyte present in the sample?" and, "How reliable are the quantitative results?" There is a song by the J. Geils Band from about 25 years ago, where the refrain is, "First, I look at the purse," meaning that the most basic evaluator of the situation is how much is it worth. I find that this philosophy accords well with many situations in life, including data review. This chapter is my approach to - and method of - organic data evaluation, regardless of whether the terminology is evaluation, verification, validation, or interpretation, or whether accuracy or bias have been determined. Analysis is expensive, and the bottom line has got to be, "Can I recover any reliable results from the money I spent?"

The ideal situation for assessing results on a sample would be to have: 1) a complete matrix spike recovery list for the sample, 2) surrogate recoveries for the sample and the matrix spike, 3) the results of analysis of the sample in duplicate, and 4) some background historical information of the sample site. A blank is nice to have, although the information gleaned from the blank is also available in the matrix spike and the duplicate sample analysis. All too often the sample-specific matrix spike is not available, and the evaluation must be performed on the surrogates and the chromatogram. This is not too bad since there is plenty of evaluation information from these two areas. We'll concentrate on them.

The chromatogram of the sample is the place to start. If the chromatogram displays obvious non-differentiated interference from co-extracted materials, evidenced by small or large hump-o-grams, note the retention times covered by the hump(s). Or the chromatogram may display one or more massive-well defined peaks as interference. Figure 8-1 displays such an interference. Again note the retention times covered by the interference. Another situation is the chromatogram that displays hundreds of small- and medium sized peaks due to co-extracted materials. Again note the retention times covered by the interferences. Not uncommon is the chromatogram that displays a flat baseline with only the surrogates (and internal standards if they are used in the analysis) as defined peaks.

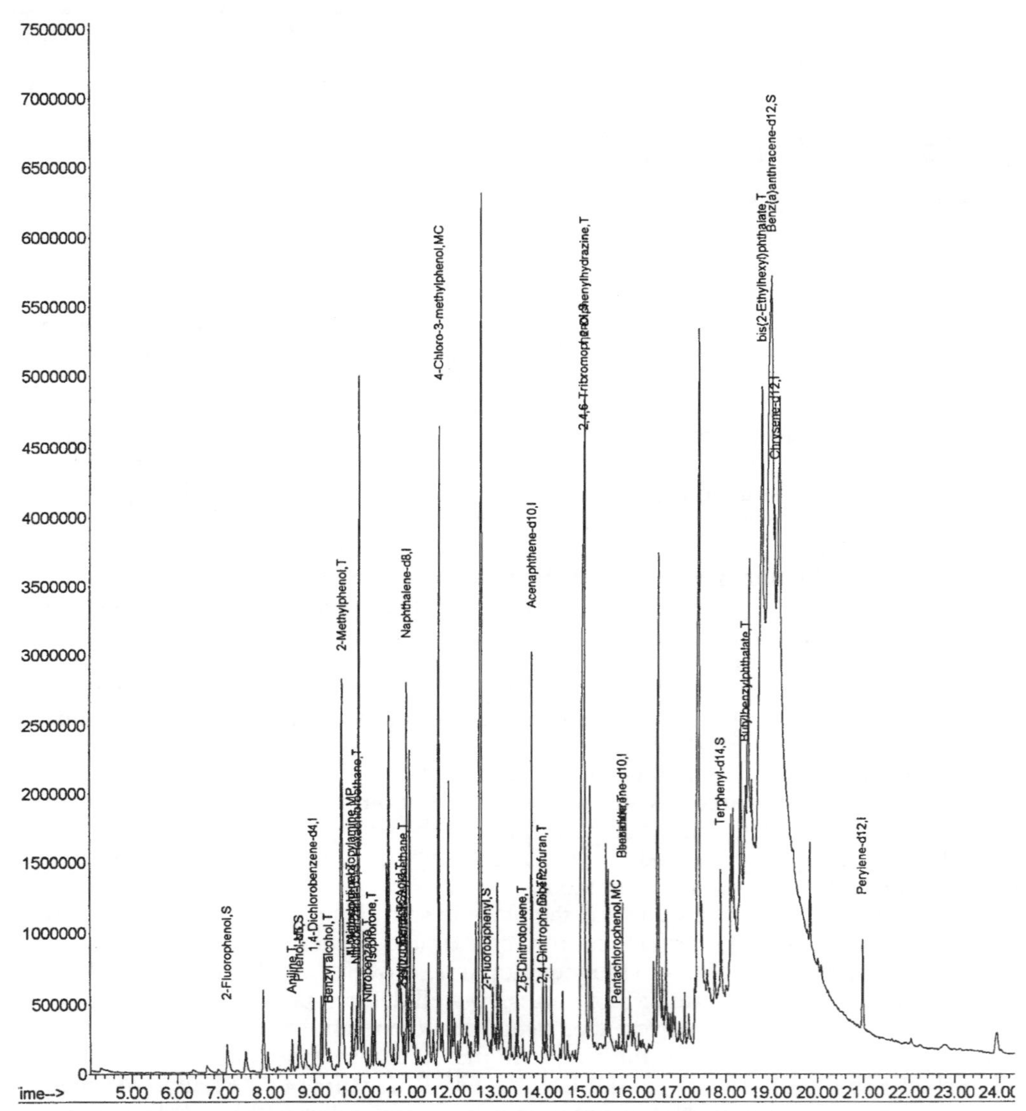

Figure 8-1. Chromatogram exhibiting a significant interference.

When attempting to obtain the maximum amount of usable information from a chromatographic analysis, particularly when interferences are present, an essential set of information is the order of elution of the target analytes from the column. This elution order table should include the internal standards and the surrogates. The most accurate elution order is that actually generated on the instruments used for the analysis. A calibration or continuing calibration summary often lists retention times for the analytes, but frequently does not include the surrogates and internal standards. Representative elution orders for BNA, VOA, chlorinated pesticides, chlorinated acid herbicides, and nitrogen-phosphorus pesticides on commonly encountered columns are presented in Tables 8-1 through 8-5.

Table 8-1. VOC target analyte (EPA Method 8260B) elution orders.

Target Analyte	Retention Time min	PQL μg/L
Dichlorodifluoromethane	1.24	3
Chloromethane	1.39	2
Vinyl chloride	1.48	1
Bromomethane	1.75	10
Chloroethane	1.84	2
Trichlorofluoromethane	2.08	3
Ethanol	2.09	250
Diethyl ether	2.39	5
Acrolein	2.51	25
1,1-Dichloroethene	2.60	0.12
Acetone	2.68	25
Acetonitrile	2.69	50
Iodomethane	2.75	2
Carbon disulfide	2.81	10
Allyl chloride	3.03	5
Methylene chloride	3.19	2
Acrylonitrile	3.54	20
trans-1,2-Dichloroethene	3.54	2
Methyl-*t*-butyl ether	3.58	5
1,1-Dichloroethane	4.18	2
Vinyl acetate	4.32	2
2,2-Dichloropropane	5.02	10
cis-1,2-Dichloroethene	5.05	2
2-Butanone	5.11	10
Propionitrile	5.19	50
Methyl acrylate	5.26	5
Bromochloromethane	5.37	2
Methacrylonitrile	5.37	50
Tetrahydrofuran	5.43	5
Chloroform	5.53	2
1,1,1-Trichloroethane	5.71	2
Dibromofluoromethane (Surr)	5.74	-
1-Chlorobutane	5.88	10
Isobutanol	5.90	250
Carbon tetrachloride	5.92	2
1,1-Dichloropropene	5.94	2
1,2-Dichloroethane-d_4 (Surr)	6.14	-
Benzene	6.20	1
1,2-Dichloroethane	6.24	2
Fluorobenzene (IS)	6.57	-
1,4-Difluorobenzene (IS)	6.74	-
Trichloroethene	7.03	2
1,4-Dioxane	7.15	100
1,2-Dichloropropane	7.30	2
Dibromomethane	7.44	2

Table 8-1. VOC target analyte (EPA Method 8260B) elution orders, *continued*

Target Analyte	Retention Time min	PQL µg/L
Methylmethacrylate	7.52	10
Bromodichloromethane	7.66	10
2-Nitropropane	7.94	20
2-Chloroethylvinyl ether	8.06	2
cis-1,3-Dichloropropene	8.20	2
4-Methyl-2-pentanone	8.41	5
Toluene-d_8 (Surr)	8.49	-
Toluene	8.57	2
trans-1,3-Dichloropropene	8.86	2
Ethyl methacrylate	9.01	2
1,1,2-Trichloroethane	9.07	2
Tetrachloroethene	9.20	2
1,3-Dichloropropane	9.25	2
2-Hexanone	9.39	10
Dibromochloromethane	9.51	1
1,2-Dibromoethane	9.61	5
Chlorobenzene-d_5 (IS)	10.17	-
Chlorobenzene	10.20	2
1,1,1,2-Tetrachloroethane	10.31	1
Ethylbenzene	10.35	2
p+m-Xylene	10.48	3
o-Xylene	10.94	2
Styrene	10.96	2
Bromoform	11.16	10
Isopropylbenzene	11.38	10
Bromofluorobenzene (Surr)	11.54	-
Bromobenzene	11.70	2
1,1,2,2-Tetrachloroethane	11.75	2
1,2,3-Trichloropropane	11.78	3
1,4-Dichloro-2-butene	11.81	5
n-Propylbenzene	11.86	2
2-Chlorotoluene	11.94	2
4-Chlorotoluene	12.07	2
1,3,5-Trimethylbenzene	12.07	2
p-Isopropyltoluene	12.45	2
Pentachloroethane	12.45	10
tert-Butylbenzene	12.45	2
1,2,4-Trimethylbenzene	12.50	2
sec-Butylbenzene	12.70	2
1,3-Dichlorobenzene	12.81	2
1,4-Dichlorobenzene-d_4 (IS)	12.89	-
1,4-Dichlorobenzene	12.92	2
1,2-Dichlorobenzene	13.35	2
n-Butylbenzene	13.36	2
Hexachloroethane	13.64	5

Table 8-1. VOC target analyte (EPA Method 8260B) elution orders, *continued*

Target Analyte	Retention Time min	PQL µg/L
1,2-Dibromo-3-chloropropane	14.27	10
Nitrobenzene	14.27	20
1,2,4-Trichlorobenzene	15.24	2
Hexachlorobutadiene	15.46	2
Naphthalene	15.52	5
1,2,3-Trichlorobenzene	15.81	2

Table 8-2. BNA target analyte (EPA Method 8270C) elution order.[1]

Target Analyte	Retention time	EQL µg/L
Pyridine	3.85	50
N-Nitrosodimethylamine	3.98	10
2-Picoline	5.41	10
N-Nitrosomethylethylamine	5.81	20
Methyl methanesulfonate	6.45	20
2-Fluorophenol (Surr)	6.63	-
N-Nitrosodiethylamine	7.09	10
Ethyl methanesulfonate	7.69	20
Aniline	8.31	10
Phenol-d_5 (Surr)	8.32	-
Phenol	8.35	10
Bis(2-chloroethyl)ether	8.43	10
2-Chlorophenol	8.45	10
1,3-Dichlorobenzene	8.66	10
1,4-Dichlorobenzene-d_4 (IS)	8.73	-
1,4-Dichlorobenzene	8.76	10
1,2-Dichlorobenzene	9.06	10
Benzyl alcohol	9.08	10
2-Methylphenol	9.30	10
Bis(2-chloroisopropyl)ether	9.31	10
Acetophenone	9.50	10
Hexachloroethane	9.55	10
Pentachloroethane	9.55	10
3-Methylphenol + 4-Methylphenol	9.57	10
N-Nitrosopyrrolidine	9.58	20
o-Toluidine	9.58	20
N-Nitroso-di-*n*-propylamine	9.60	10
N-Nitrosomorpholine	9.61	10
Nitrobenzene-d_5 (Surr)	9.76	-

[1] 30 m x 0.25 mm id Restek RTX-5, 1 mL/min constant flow, 40 °C for 4 min, followed by 15 °C/min to 160 °C, then 20 °C/min to 310 °C for 6.5 min; total run time 26.5 min.

Table 8-2. BNA target analyte (EPA Method 8270C) elution order[2], *continued*

Target Analyte	Retention time	EQL µg/L
Nitrobenzene	9.79	10
N-Nitrosopiperidine	10.05	20
Isophorone	10.16	10
2-Nitrophenol	10.30	10
2,4-Dimethylphenol	10.41	10
Bis(2-chloroethoxy)methane	10.56	10
O,O,O-Triethylphosphorothioate	10.57	20
2,4-Dichlorophenol	10.67	10
1,2,4-Trichlorobenzene	10.79	10
Naphthalene-d_8 (IS)	10.86	-
Benzoic acid	10.86	50
Naphthalene	10.90	10
a,a-Dimethylphenethylamine	10.93	30
2,6-Dichlorophenol	11.05	10
4-Chloroaniline	11.07	10
Hexachloropropene	11.08	30
Hexachlorobutadiene	11.19	10
N-Nitrosodibutylamine	11.66	10
1,4-Phenylenediamine	11.68	10
4-Chloro-3-methylphenol	11.87	20
Isosafrole	11.93	10
2-Methylnaphthalene	12.03	10
1,2,4,5-Tetrachlorobenzene	12.37	10
Hexachlorocyclopentadiene	12.38	20
2,4,6-Trichlorophenol	12.54	10
2,4,5-Trichlorophenol	12.60	10
2-Fluorobiphenyl (Surr)	12.67	-
Safrole	12.74	50
2-Chloronaphthalene	12.80	10
2-Nitroaniline	13.07	50
1,4-Naphthoquinone	13.12	10
Dimethyl phthalate	13.40	10
Acenaphthylene	13.44	10
2,6-Dinitrotoluene	13.50	20
Acenaphthene-d_{10} (IS)	13.66	-
3-Nitroaniline	13.70	50
Acenaphthene	13.72	10
2,4-Dinitrophenol	13.82	50
4-Nitrophenol	13.94	50

[2] 30 m x 0.25 mm id Restek RTX-5, 1 mL/min constant flow, 40 °C for 4 min, followed by 15 °C/min to 160 °C, then 20 °C/min to 310 °C for 6.5 min; total run time 26.5 min.

Table 8-2. BNA target analyte (EPA Method 8270C) elution order[3], *continued*

Target Analyte	Retention time	EQL µg/L
Dibenzofuran	13.95	10
1,3-Dinitrobenzene	13.95	10
4-Bromophenyl phenyl ether	13.98	10
Pentachlorobenzene	13.98	10
2,4-Dinitrotoluene	14.07	20
2-Naphthylamine	14.07	10
1-Naphthylamine	14.09	10
4-Nitroquinoline-1-oxide	14.11	10
2,3,4,6-Tetrachlorophenol	14.19	10
Diethyl phthalate	14.40	10
Fluorene	14.43	10
4-Chlorophenyl phenyl ether	14.44	10
Thionazine	14.50	20
4,6-Dinitro-2-methylphenol	14.59	50
4-Nitroaniline	14.60	50
Diphenylamine	14.64	10
5-Nitro-*o*-toluidine	14.64	10
N-Nitrosodiphenylamine	14.64	10
Diphenyl hydrazine	14.65	10
2,4,6-Tribromophenol (Surr)	14.79	-
Sulfotepp	14.99	20
Phorate	15.08	20
Diallate, total	15.15	10
1,3,5-Trinitrobenzene	15.15	10
Phenacetin	15.17	100
Hexachlorobenzene	15.29	10
Dimethoate	15.36	20
4-Aminobiphenyl	15.48	10
Pentachlorophenol	15.53	50
Pronamide	15.61	10
Pentachloronitrobenzene	15.65	10
Phenanthrene-d_{10} (IS)	15.69	-
Phenanthrene	15.73	10
Disulfoton	15.76	10
Dinoseb	15.78	20
Anthracene	15.79	10
Methyl parathion	16.26	10
Di-*n*-butyl phthalate	16.51	10
Parathion	16.75	10

[3] 30 m x 0.25 mm id Restek RTX-5, 1 mL/min constant flow, 40 °C for 4 min, followed by 15 °C/min to 160 °C, then 20 °C/min to 310 °C for 6.5 min; total run time 26.5 min.

Table 8-2. BNA target analyte (EPA Method 8270C) elution order[4], *continued*

Target Analyte	Retention time	EQL µg/L
Methapyrilene	16.88	10
Isodrin	17.06	20
Fluoranthene	17.25	10
Benzidine	17.40	80
Pyrene	17.53	10
Terphenyl-d_{14} (Surr)	17.73	-
Aramite	17.83	20
Dimethylaminoazobenzene	17.93	10
Chlorobenzilate	18.00	50
Kepone	18.35	20
3,3'-Dimethylbenzidine	18.35	10
Famphur	18.36	20
Butyl benzyl phthalate	18.38	10
2-Acetylaminofluorene	18.71	10
Methoxychlor	18.97	10
Benz(a)anthracene	19.02	10
3,3'-Dichlorobenzidine	19.03	10
Chrysene-d_{12} (IS)	19.05	-
Bis(2-ethylhexyl)phthalate	19.08	10
Chrysene	19.09	10
Di-*n*-octyl phthalate	19.80	10
Benzo(b)fluoranthene	20.38	10
7,12-Dimethylbenz(a)anthracene	20.39	10
Benzo(k)fluoranthene	20.41	10
Hexachlorophene	20.65	100
Benzo(a)pyrene	20.81	10
Perylene-d_{12} (IS)	20.88	-
3-Methylcholanthrene	21.36	10
Indeno(1,2,3-cd)pyrene	22.70	10
Dibenz(a,h)anthracene	22.72	10
Benzo(g,h,i)perylene	23.22	10

Table 8-3. Chlorinated pesticide target analyte (EPA Method 8081A) elution order.

Target Analyte	RT DB 608	RT DB 1701
Trifluralin	3.49	6.24
TCMX (surr)	4.25	4.86
Hexachlorobenzene	5.73	6.55
α-BHC	6.14	7.50

[4] 30 m x 0.25 mm id Restek RTX-5, 1 mL/min constant flow, 40 °C for 4 min, followed by 15 °C/min to 160 °C, then 20 °C/min to 310 °C for 6.5 min; total run time 26.5 min.

Table 8-3. Chlorinated pesticide target analyte (EPA Method 8081A) elution order, *continued*

Target Analyte	RT DB 608	RT DB 1701
γ-BHC	7.40	8.84
β-BHC	7.63	11.51
Heptachlor	8.51	9.58
δ-BHC	8.92	12.46
Aldrin	9.69	10.95
Isodrin	11.54	12.31
Heptachlor epoxide	11.84	13.14
γ-Chlordane	12.47	14.40
α-Chlordane	13.13	14.62
Endosulfan I	13.20	14.21
4,4'-DDE	14.21	15.00
Dieldrin	14.46	15.65
Endrin	15.95	16.46
Toxaphene (main peak)	16.29	17.34
4,4'-DDD	16.51	18.17
Endosulfan II	16.71	18.41
4,4'-DDT	17.75	18.84
Endrin aldehyde	18.10	20.11
Endosulfan sulfate	18.61	21.14
DCB (surr)	19.90	21.60
4,4'-Methoxychlor	21.49	21.35
Endrin ketone	21.81	23.10
Octachloronaphthalene	28.15	27.24
DBC (surr)	28.22	27.40

Table 8-4. Chlorinated acid herbicide target analyte (EPA Method 8151A) elution order

Target Analyte	RT 608	RT 1701	PQL ug/L
Dalapon	3.76	4.29	25
DCAA (surr)	9.46	9.77	-
MCPP	9.68	10.28	1000
Dicamba	9.86	10.11	1
MCPA	10.20	10.69	1000
Dichloroprop	10.61	11.19	10
2,4-D	11.26	11.74	10
Pentachlorophenol	12.1	12.2	10
2,4,5-TP	12.28	12.93	1
2,4,5-T	13.09	13.63	1
Dinoseb	13.26	15.40	5
2,4-DB	13.75	14.32	10
Picloram	14.70	16.4	10

Table 8-5. Nitrogen-phosphorus pesticide target analyte (EPA method 8141A) elution order.

Target Analyte	RT 608	DB-5	PQL mg/L
Dichlorovos	4.76	4.54	1
1-Bromo-2-nitrobenzene (IS)	5.85	-	-
Mevinphos	6.98	7.76	1
Demeton-O	9.11	11.40	1
Ethoprop	9.36	11.84	1
Tributylphosphate (Surr)	9.40	11.90	-
Sulfotepp	9.88	12.89	1
Phorate	10.02	13.08	1
Naled	10.02	12.46	1
Demeton-S	10.36	13.70	1
Dimethoate	10.38	13.90	1
Diazinon	11.13	15.19	1
Disulfoton	11.20	15.35	1
Parathion Methyl	12.06	17.08	1
Ronnel	12.38	17.59	1
Merphos	13.01	21.37	1
Trichloronate	13.01	19.28	1
Parathion	13.01	18.86	1
Chlorpyrifos	13.29	18.81	1
Fenthion	14.32	18.81	1
Tokuthion	14.67	22.09	1
Stirophos	14.76	21.96	1
Bolstar	15.46	24.28	1
Fensulfothion	15.85	23.55	1
Famfur	15.96	24.67	1
Triphenylphosphate (Surr)	16.52	25.70	-
Azinphos Methyl	17.65	27.37	1
Coumaphos	18.74	28.95	1

These tables are most useful when the retention time boundaries for an obvious chromatographic interference can be established. In Figure 8-1 boundaries should be set from 17.0 minutes to 21.0 minutes. Analyte identification and quantitation is expected to be affected by the interference; however, instead of flagging all the results in the analysis, only those target analytes eluting within the boundaries of the interference can be qualified. Another use of the table is when only a few of the surrogates or internal standards in the analysis are exhibiting recoveries that are out of the normal. Target analytes associated with just those surrogates or internal standards can be flagged.

The qualitative information obtained from the chromatogram serves as a guide as to where to go next, and next is most often the surrogate recoveries.

A. Surrogates

Surrogates are evaluated by acceptance ranges that are generated from the laboratory's own historical performance of the analysis. These ranges must be specific for the type of sample and the sample extraction procedure. As an example, the surrogate acceptance ranges for the BNA surrogates from a Method 3510/8270 procedure performed on water samples are presented in Table 8-6. The data set for the calculation included laboratory control samples, blank samples, and over 250 different sources of environmental water samples (both surface and ground water) for a total of over 4000 sets of surrogate recoveries. For comparison the literature values for the same extract concentration methodology and sample type from the Contract Laboratory Program Statement of Work (OLM04.2) and the U.S. Air Force Quality Assurance Project Plan (Version 3.1) are included. It should be obvious that if a more efficient type of sample preparation is used, such as use of continuous liquid-liquid extraction (EPA Method 3520) with in-device sample concentration rather than the separatory funnel and Kuderna-Danish concentration described in Method 3510, that different (higher) surrogate recovery ranges would be obtained.

Table 8-6. Historical surrogate recovery limits for water samples processed by EPA Method 3510/8270 with K-D concentration from a laboratory as compared to literature values for the sample matrix and methods.

Surrogate	CLP	AFCEE	Laboratory
2- Fluorophenol (FP)	21-110	25-125	21-100
Phenol-d_5 (Pd_5)	10-110	25-125	10-94
Nitrobenzene-d_5 (NBd_5)	35-114	32-125	35-114
2-Fluorobiphenyl (2-FBP)	43-116	43-125	43-116
2,4,6-Tribromophenol (246-TBP)	10-123	25-134	10-123
Terphenyl-d_{14} (Td_{14})	33-141	42-126	33-141

Table 8-7. Characteristic 8270 surrogate recovery patterns for different situations.

Normal recoveries					
FP	Pd_5	NBd_5	2-FBP	246-TBP	Td_{14}
30	29	54	55	85	91
Low/missing acid pattern					
FP	Pd_5	NBd_5	2-FBP	246-TBP	Td_{14}
1	0	54	55	3	91
Low/missing B/N pattern					
FP	Pd_5	NBd_5	2-FBP	246-TBP	Td_{14}
30	29	1	0	85	0
Slight over concentration					
FP	Pd_5	NBd_5	2-FBP	246-TBP	Td_{14}
1	0	35	40	85	91
Severe over-concentration					
FP	Pd_5	NBd_5	2-FBP	246-TBP	Td_{14}
0	0	2	1	85	91
Column active-site pattern					
FP	Pd_5	NBd_5	2-FBP	246-TBP	Td_{14}
1	3	3	56	0	85
Localized co-eluting matrix interference					
FP	Pd_5	NBd_5	2-FBP	246-TBP	Td_{14}
30	28	0	56	85	91
General co-eluting matrix interference					
FP	Pd_5	NBd_5	2-FBP	246-TBP	Td_{14}
30	29	0	0	0	85

As discussed in chapter 1, the most informational situation is when more than one surrogate has been used in the analysis. Otherwise the deviant behavior of the single surrogate is frequently assumed to be displayed also by all the analytes in the determination. Data that may be completely reliable is flagged as questionable.

Table 8-7 presents a number of possible scenarios for surrogate recoveries of the BNA surrogates of the EPA GC-MS Method 8270C. Each of these patterns of recoveries can be indicative of certain problems in the analysis. Once a problem area has been identified, the qualifying of results for the limited number of target analytes that are subject to the same problem can be performed in a very selective fashion. Several examples can serve to illustrate these type evaluations.

In the first example, a sample displays the surrogate recovery pattern similar to the second entry in Table 8-7. In this case only the acidic surrogates are exhibiting low recoveries. This may be due to the extraction of the acidified sample not being performed, or perhaps there is a separate GC-MS analysis of the acid fraction. There also could be a particular on-column interference left over from a previous sample that prevents determination of acidic components. At any rate these results would indicate that any of the target analytes that exhibit acidic properties, such as the phenols and any organic acids that are reported from this GC-MS run, are not reliable, particularly results that indicate the target analyte was not detected. On the other hand there is no evidence in this surrogate recovery pattern that there is any problem with the analysis of the base/neutral target analytes, and they should not be qualified. Therefore the appropriate action is to flag acidic target analytes as unreliable (R) if they are reported as undetected. It's a low probability that there would be reported hits for acidic target analytes under these situations, but if they were reported as present, the appropriate flag should indicate these results are estimated (J).

The second example concerns the surrogate pattern in the fourth row of Table 8-7. The cause of this characteristic pattern is an over zealous concentration of the sample extract. The target analytes that would be expected to exhibit similarly low recoveries are those that are the early eluters in the chromatogram. With reference to Table 8-2, the retention times for 2-fluorophenol and phenol-d_5 are 6.63 and 8.32 minutes respectively. The next eluting surrogate is nitrobenzene-d_5 at 9.76 minutes, which is recovered within the acceptable range. Target analytes that elute prior to 8.32 minutes definitely have to be qualified, and the likely scenario is that some of the later eluting target analytes, say up to half way between the retention times of phenol-d_5 and nitrobenzene-d_5, around 9.0 minutes also need to be qualified. The applicable qualification is that affected analytes that are detected are probably biased low in the reported result, meaning the results are estimated (J). Analytes reported as undetected may in actuality be present in the sample; thus the data are unreliable (R).

The third example concerns the sixth pattern of surrogate recoveries presented in Table 8-7, the localized matrix interference example. Here only one surrogate recovery is out of the acceptance range, and although the example illustrates a low recovery, samples can just as easily exhibit high recoveries. If the surrogate falls within the retention time boundaries of a visible interference in the chromatogram, the data reviewer is perfectly justified in qualifying results on all target analytes that also elute within those boundaries. The question is how to perform the qualification. The choices are either that all results are estimated (J), or that all results are unreliable (R). There is really not sufficient information to say that, "because the surrogate recovery is high, all associated target analytes should be viewed as biased high," or, "because the surrogate recovery is low, all associated target analyte recoveries are similarly biased low." This is particularly true in the case of the elevated surrogate recovery, because the reason for the elevation may be due to either fortuitous presence of the surrogate quantitation ion in the interference, or it may be due to significantly low recovery of the internal standard. Unless the reason can be pinpointed, it is difficult to rationalize selective qualifications of target analytes based on whether the analyte is reported as detected or not detected. The same thought process is applicable when the sample exhibits widespread matrix interference, as represented by the last entry in Table 8-7.

Another situation that is encountered is when sample extracts have been diluted and re-analyzed. This occurs in one of two cases, one of which is acceptable and straight-forward in evaluation. The other presents many interpretational problems.

The acceptable case (Case I in Table 8-8) is encountered when one or more detected target analytes are over the calibration range. The analysis of the undiluted sample exhibits acceptable surrogate recoveries, indicating that the sample preparation was under control and that interferences present in the sample are not significant. All target analyte results that are reported from the undiluted sample are reliable. Subsequent dilution and re-analysis of the sample extract to obtain defensible quantitation of some target analytes frequently dilutes out the surrogates, giving results like those presented in the diluted sample of Case I. No data can rationally be qualified in this situation. The acceptable surrogate recoveries from the undiluted sample have already served the purpose of demonstrating the analysis is reliable.

Table 8-8. Common surrogate patterns from diluted samples.

CASE I					
Undiluted Sample					
FP	Pd_5	NBd_5	2-FBP	246-TBP	Td_{14}
30	29	54	55	85	91
Diluted Sample					
FP	Pd_5	NBd_5	2-FBP	246-TBP	Td_{14}
0	0	0	0	1	2
CASE II					
Undiluted Sample					
FP	Pd_5	NBd_5	2-FBP	246-TBP	Td_{14}
0	0	0	1	0	3
Diluted Sample					
FP	Pd_5	NBd_5	2-FBP	246-TBP	Td_{14}
0	0	0	0	1	2
CASE III					
Undiluted Sample					
FP	Pd_5	NBd_5	2-FBP	246-TBP	Td_{14}
0	0	0	1	0	3
Diluted Sample					
FP	Pd_5	NBd_5	2-FBP	246-TBP	Td_{14}
30	29	54	55	85	91

Laboratories that make a full and honest disclosure of all quality control results from every analysis perhaps do themselves a disservice in a situation like Case I. Too much information can sometimes muddy the interpretation and confuse the client rather than provide clarification.

Case II from Table 8-8; however, presents major interpretational problems to the data user. Case II occurs most frequently when the sample exhibits major analytical interferences, and the laboratory has attempted to circumvent the problem by the unconscionable expedient of dilute and re-shoot. My response as a data reviewer to a

Case II situation is to append a flag of unreliable (R) to all the sample results. There is no evidence in this case that the sample has been properly prepared or handled carefully. Granted there may be the rare situation where the data needs are only qualitative; all that is needed is identification of compounds present in the sample in significant abundance. In this situation the results may be usable. But the more frequent situation requires a defensible analysis and these data do not meet these needs in any respect. The appropriate behavior by the laboratory should be to either re-process a smaller portion of the sample, or to employ relevant clean-up procedures to remove the interferences.

There are situations when, due to lack of sample, clean-up must be performed on the original sample extract. This can result in loss of surrogates, particularly when the surrogates are not compatible with the clean-up. An example of this is encountered in PAH analysis by EPA Method 8270C. The original extraction may contain significant interferences from saturated hydrocarbon fuels. A subsequent clean-up of the extract with silica gel (described in Chapter 1) is going to result in significant loss of the nitrobenzene-d_5 and all the phenolic surrogates. This is an understandable and unavoidable loss of surrogate, and it will be revealed in the report of surrogate recoveries. However, all is not lost. There are still two surrogates that should be present, 2-fluorobiphenyl and terphenyl-d_{14}, and they can be used for data evaluation. But even more fortuitous, the extract often contains the internal standards when it is subjected to the PAH clean-up. The area counts of the internal standards in the initial analysis and the re-analysis after clean-up can be compared to evaluate recovery. Further, the presence of the internal standards, which are deuterated PAH, will internally compensate for slightly low recoveries in the clean-up. The Case Narrative for the analysis should mention any or all of these occurrences.

There is another situation that is represented by the Case III surrogate recoveries in Table 8-8. It occurs rarely, but it does occur. The interferences in the sample extract overload the ability of the capillary column to adequately separate sample components. Symptoms of this are distorted peak shapes and shifted retention times. Dilution of the sample, normally by a factor of no more than 5, can result in reduction of the on-column overload, so that the column can perform properly, generating acceptable surrogate identification and quantitation. This occurrence will still result in raised detection limits for all target analytes by an amount equal to the dilution factor. If the higher detection limits are still usable by the client, the analysis can be viewed as satisfactory. Otherwise a re-extraction or employment of an appropriate clean-up needs to be performed.

B. Matrix Spikes and Laboratory Control Samples

The interpretational utility of the matrix spikes (MS) and laboratory control samples (LCS) changes depending on the analyte contents of the spikes and whether a duplicate (MSD or LCSD) was performed at the same time. Let's look at the duplicates first.

First, anytime a duplicate is performed during an analysis, be it a sample duplicate, a matrix spike duplicate, a laboratory control sample duplicate, or even a sample dilution re-analysis, an opportunity is presented to evaluate the level of laboratory contamination. This is particularly significant with regard to the common contaminants discussed in Chapter 7. In the case of sample interference exhibited by a sample chosen for a matrix spike, the MS and MSD chromatograms offer an evaluation of the possibility that the sample interference is laboratory generated rather than being an intrinsic property of the sample.

Second, performance of duplicates allows evaluation of precision. As a rule of thumb, laboratories that use automated equipment for sample preparation are going to offer better precision than largely manual operations. Granted there are very few facilities that are completely automated, including robotic sample manipulation; however, steps the laboratory has taken to insure that exactly the same procedures are used for every sample, will generally show up as improved analytical precision. Note that improved accuracy is not necessarily a given when less hands-on manipulation of samples is used. The philosophy of most probable outcome (Ockham's Razor) suggests that when a machine is not operating properly it tends to not operate properly on every single sample.

Precision can be evaluated in either a qualitative sense or quantitatively. The quantitative approach normally consists of a relative percent difference (RPD) calculation and requires that the compounds evaluated be calibrated. The equation for RPD is simple:

$$RPD = \frac{2|A-B|}{A+B} \times 100$$

where A and B are the analyte results from the individual determinations. However, RPD as a metric of precision is biased toward high numerical values of the results and can give skewed evaluations. For quantities of analytes close (within a factor of 5) to the reporting limit of the test, a more reliable yardstick is to add and subtract the reporting limit from the results, then perform a comparison to see if the ranges overlap. If they do then the results are considered to be the same.

Laboratory contamination evaluations based on performance of MS and LCSD duplicates is going to be much easier when there are only a few added target analytes to the samples. When full analyte list matrix spikes and laboratory controls have been performed, the information about laboratory contamination is still contained in the results and chromatograms, it's just more tedious to pick it out and only occasionally worth the effort. But in cases where one or more of the common lab contaminants are reported as hits in samples, the effort needs to be expended.

The MS, MSD, LCS, and LCSD can be evaluated as a group, particularly when the same compounds and similar spike amounts are present in all of them. Random low or high recoveries for a target analyte can be spotted and disregarded if one control is out but the other three are with in the acceptance ranges. On the other hand, three of the controls may exhibit low (or high) recoveries for a target analyte while the last recovery is within range. The evaluation should be that the "good" result is a fluke, and the other three results are the norm for the analysis, rather than the analyst feeling relief that at least one of the results passes, and the run is acceptable.

The matrix spike is reliable as a general indicator of sample matrix effects in the analysis only when the samples are very similar. Examples of batches that consist of samples that could be considered very similar are 20 fly ash samples, or 20 groundwater samples from the same monitoring field, or 20 soil samples from a single sampling grid. In these type situations, the results on a matrix spike can sometimes be generalized to all the samples in the batch. Evaluations of this type are reinforced when more than one compound in the spike solution is demonstrating deviant behavior, such as, all the acids are low, or all the spike compounds eluting between 14.0 and 17.0 minutes are low. The same sub-groupings of compounds that were described above in relation to the surrogate evaluations, are applicable in MS/LCS evaluations. Corroboration of the effect by the behavior of the surrogates is expected, and in fact the order of events is most frequently reversed during an actual data evaluation event: the low (or high) recoveries are first noted in the surrogates, then corroborated in the behavior of the matrix spike compounds.

The more commonly encountered batch contains samples from diverse sources. Examples are 20 wastewater effluent samples from 15 different effluent streams, or a batch containing 20 soil, sediment, and sludge samples from different locations. Even when the batch contains 20 soil samples that were collected during one sampling event on one site, the samples may have been obtained in a variety of microenvironments such that a range of soil types and geologies are represented. In these situations the generalization of the matrix spike results to all the samples is on a very shaky foundation[5]. The quality control is essentially non-relevant for sample result evaluations, unless your particular sample was fortuitously chosen for the matrix spike.

The acceptance ranges for matrix spike and laboratory control compounds must be derived in the laboratory from historical data. As described in Chapter 5, there should be separate derivations for the two, and the laboratory control ranges should be smaller than the associated matrix spike ranges. Further, there should be a set of ranges derived from different matrices, at least one for water samples, another for soils, and others for sludges (biosolids) and TCLP extracts.

C. Duplicates

There are several different types of duplicates that can be analyzed. A laboratory duplicate for semivolatile analytes is where the analyst mixes a sample thoroughly, then weighs out two portions of the sample, and processes the portions side-by-side through the entire analytical protocol. What is done to one portion gets done exactly in the same way to the other portion. The results from duplicates treated in this fashion should be very, very similar. The chromatograms should be almost identical. The same set of target analyte hits should be present in both portions and in the similar amounts. The surrogate recoveries should be very similar. Internal standard areas should be very similar. Variations in the results should be due only to laboratory random error, quantitatively, for compounds present in amounts well above the reporting limit, on the same order of magnitude as the precision exhibited by laboratory control or matrix spike duplicates. Significant differences in the sample results are a sign that the causes should be investigated and determined. The most frequent cause of variation is insufficient sample mixing prior to taking the duplicate portions.

When water samples are being analyzed for semivolatiles, normally the contents of a single sample container is one sample. If duplicates are to be analyzed, duplicate samples should be collected in the field. This automatically introduces more variation into the results, although water sources in the field should be inherently more homogenous than soil samples. If duplicate samples are not available in separate containers, and the laboratory has to split the contents of a single container, this introduces significant variation to the process. Organic analytes are generally not homogeneously distributed in a bottle of sample, instead they tend to be concentrated on the bottle walls and in a film at the top of the sample. Thus the first and last sample portions decanted from a bottle tend to be different, even when the contents have been vigorously shaken.

When volatile analytes are being determined, duplicates of water samples tend to be much more similar than duplicates of soil samples. The current protocol for soil sampling (EPA Method 5035) allows only a single analysis from each sample container. Laboratory duplicates are actually field sampling duplicates, taken as co-located samples rather than as two portions from a well-mixed sample. The process has considerable built-in variation, and the results often reflect this.

The interpretation of laboratory duplicate results is considerably easier than field duplicates. There are two types samples that are collected and frequently labeled as

---

[5] The exception to this rule is batch analysis of TCLP extracts. As a generality these extracts, even when prepared from a variety of samples, are very similar in behavior.

"field duplicates." The first is where the sampler insures that the samples are as similar to each other as possible. This requires obtaining a sample, then vigorously and thoroughly mixing it prior to containerizing two portions of the sample. This process generates two fairly similar samples, which can be treated as - and correctly termed - "field duplicates." On the other hand if the sampler scoops one portion of soil from a site and puts it in a container, then scoops out a second portion of soil and puts it in a second container, these are not duplicates. Instead they are co-located samples and will exhibit more variation than a well-mixed field duplicate. Most samples that are received by labs and are labeled as field duplicates are actually co-located samples.

A correctly prepared field duplicate can be very useful as a quality control check on the laboratory. The duplicate portions are labeled with different sampling codes and introduced at separated points into the flow of samples being sent to the lab. The results on the duplicates can then be compared for variance much the same as the laboratory duplicate.

A frequent interpretational problem that tends to arise is when one of the samples in a duplicate pair exhibits a hit for a target analyte that is slightly above the reporting limit while the analyte in the other sample is reported as less than the reporting limit. This problem is seen more often in situations where the reporting limit for the data has been forced down to levels that are barely within the capabilities of the instrument and the analytical procedure. This occurrence creates a significant problem for all computers and some persons. The common characteristic of computers and these people is that they can only evaluate data in terms of black & white, good & bad, acceptable & failing. The root of the problem is that values under the reporting limit are equated with zero. The art of interpretation is the ability to recognize and accept shades of gray rather than insist that everything must be classified as yes or no.

CHAPTER 9

Other Data Assessments

So far, I have intentionally not mentioned regulatory method compliance requirements. These aspects of data generation can be characterized as legalistic imperatives that are designed to reduce the number of choices that an analyst would normally make in the performance of an analysis.

Originally there were good reasons behind these requirements. The first and foremost was to assure that all practitioners of the methods were performing them in exactly the same way so that exactly comparable monitoring data were being generated. Comparable data are important when effluents are being monitored and evaluated in a pass/fail sense against a numerical standard - the effluent discharge limit. The analytical results used to make these pass/fail determinations must be independent of whichever laboratory performed the work. The second reason was to present a complete, detailed method that could be implemented by persons possessing only a basic science education.

One of the problems with these methods, particularly the EPA 600-series methods for wastewater compliance monitoring, and also one of their great strengths, was a liberal use of the terms "shall" and "must". These terms have legal interpretations such that they indicate procedural imperatives. To ignore a shall or must means that the prescribed procedure was not followed. Unfortunately the authors of the methods did not take into account probable future scientific advancements. How could they? They were on the cutting edge of analysis and hard pressed to develop and write robust procedures that they were in the process of inventing. Looking back on these methods with over 25 years of hindsight, it is easy to suggest modifications of the procedural descriptions that could incorporate the ability to adapt new technologies into the procedure. One such major advancement is the now - widespread use of capillary columns over the packed columns that are described in the methods.

Another unfortunate aspect of the present-day use of these methods is that analysts can easily see that the methods are not up-to-date, and in many cases they know better ways to accomplish the same analytical objectives. These analysts tend to get in the habit of ignoring the required method altogether in favor of following a more current procedure. One particularly prevalent substitution is the use of the most modern version of EPA Method 8270 from the Solid Waste program instead of the required EPA Method 625. One of two consequences can stem from this substitution, depending on how it is done. The first is where the analyst reports the results as obtained using EPA Method 8270. This is simply a violation of the regulations that EPA has promulgated to implement the requirements of the Clean Water Act. The second situation is more serious in that analysts are aware that Method 625 is the legally required procedure, so they report that Method 625 was used to generate the data, but in actuality they used EPA Method 8270. This goes beyond a simple violation of regulations into the realm of fraud. Further there is the probability of generating data of lower quality[1] than if the proper method had been performed.

There is another area of method compliance that has little to no experimental basis yet is widely used as the primary gotcha by legalists and data validators. For those

[1] See R.-K. Smith, 1999, *Handbook of Environmental Analysis, 4th Edition*, Section 3, Method 625 for a detailed discussion (Genium Publishing, Schenectady, NY).

readers that are interested in obtaining money from the EPA, proposed scientific studies of sample holding times and preservation are sorely needed. Studies that have been performed to date indicate that the requirements listed in the various regulations are of no more scientific value than outright guesses.

A. Holding Times and Temperature Preservation

There are two fundamental reasons[2] behind temperature adjustment and control when samples are collected and stored for analysis. The first is to slow chemical and biological degradation of target analytes that may be in the sample. The second is to slow volatilization of target analytes from the sample, although a hermetically sealed VOA vial or the newer Encore® samplers[3] for soils reduces volatilization to a minimum. Although volatilization and chemical/biological degradation are distinct processes, the mathematical description of the rates of these processes is similar with regard to temperature dependence. The basic description of reaction rates is the Arrhenius equation[4]:

$$\text{rate} = Ae^{-E^*/RT}$$

where A is a constant specific to the reaction under consideration, E^* is the activation energy of the process, R is the gas constant, and T is the temperature. From this equation it is obvious that the higher the temperature, the faster the rate.

The temperature of 4 °C that is listed in Table II, *Required Containers, Preservation Techniques, and Holding Times*, 40 CFR 136.3 (1 July, 1997) and is widely used as the standard temperature for sample preservation, was chosen for several reasons.

1. It is sufficiently cool so that significant retardation of degradation rates occurs. The temperature effect on chemical and biological reaction is exponential, as presented above. As a rule of thumb for every 10 °C rise in temperature, the reaction rate is doubled, and for every 10 °C lessening of temperature the reaction rate is halved. Storage of a sample at 4 °C as compared to room temperature storage at 25 °C, means that degradation/volatilization occurs at approximately 1/4 the rate.
2. It is the equilibrium point of a loose ice-water slurry, and thus a bath of this temperature is easily attained. Having an ice-water slurry at the correct temperature of 4 °C does not mean that immersion of samples in the slurry instantaneously changes the temperature of the sample to be the same as the slurry. Transfer of heat from one body to another is an exponential process. There is an initial rapid cooling of the sample to within a few degrees of the slurry, then gradual approach to equilibration. The rate of equilibration is a function of the size and construction of the sample container. A glass container will equilibrate much faster than a plastic container due to the greater heat transfer rate of glass compared to plastic. Large containers will equilibrate slower than small containers due to the amount of heat that must be transferred

---

[2] USEPA, April, 1994. *Industrial User Inspection and Sampling Manual for POTW's*, EPA 831-B-94-001, pg 74.

[3] Encore® is a registered trademark of En Novative Technologies, Inc., 1241 Bellevue St. Green Bay WI 54302, 1-888-411-0757.

[4] Castellan, G.W.,1971. *Physical Chemistry, Second Edition*, Addison-Wesley Publishing Company, Reading, MA, page 746.

from the sample to the ice-water slurry. Normally a 1L glass container full of sample, initially at 25 °C, immersed in a slurry of ice-water will take 4 hours to come to exactly 4 °C. A plastic container of similar size will take longer to equilibrate.

3. It is above the freezing point of water. Solid water has less density than liquid water (this is why ice floats on water). The corollary of this phenomenon is that ice occupies more space than an equivalent weight of water. If the sample freezes, it will expand and can break the sample container. Thus freezing is not desirable due to sample container breakage; however, more on this below.

Holding times for sample analysis were established based on the assumption that if the samples are cooled and stored at 4 °C, then there will be insignificant degradation within a set period of time. Lacking data, time periods were estimated as seven days for semivolatile analytes in water, and 14 days for semivolatiles in soil and volatiles in both matrices. Results reported from recent studies on holding times demonstrate that these estimates are no more reliable than random guesses. However, they were incorporated into the regulations.

Considerable work on holding times and preservation has been done by Alan Hewett at the U.S. Army Corps of Engineers Cold Regions Research and Engineering Laboratory on volatile compounds.[5] He sealed soil spiked with known concentrations of the aromatic hydrocarbons benzene, toluene, ethylbenzene, and the xylenes and the halogenated volatile compounds trichloroethene (TCE) and tetrachloroethene (perchloroethene, PCE) in ampules, then stored the replicate ampules under a variety of conditions for varying lengths of time. For biologically active soils there was almost complete loss of the aromatic compounds after storage at 22 °C for five days; however, there was no loss of the halogenated compounds up to 21 days. If the sample was stored at 4 °C, the degradation rate was slow enough to give insignificant degradation of the aromatic compounds up to nine days, however by 14 days there was significant loss. If the samples were acidified to halt biological activity, there was insignificant loss of any compound up to 21 days even when stored at 22 °C. Other means of halting biological activity, such as addition of methanol to the sample, have been demonstrated to allow holding times of at least 28 days for both halogenated and aromatic volatile compounds.[6] Acidification and storage of water samples at 4 °C, has been shown to be suitable preservation for up to 28 days and beyond.[7]

If the sample is stored cooler than 4 °C, then the degradation rate is even further retarded, thus the sample results are unchanged for a longer period of time. For instance storage at -12 to -15 °C gives a degradation rate that is substantially less than the degradation rate at 4 °C, and the holding time for volatiles before any significant

---

[5] Hewett, A,1995. Determining volatile organic compound concentration stability in soil. *Proceedings of the Eleventh Annual Waste Testing & Quality Assurance Symposium,* 23-28 July, 1995, Washington Hilton Hotel and Towers, Washington DC.

[6] Turriff, D., R. Reitmeyer, L. Jacobs, and N. Melberg, 1995. Comparison of alternatives for sampling and storage of volatiles in soil.. *Proceedings of the Eleventh Annual Waste Testing & Quality Assurance Symposium,* 23-28 July, 1995, Washington Hilton Hotel and Towers, Washington DC.

[7] Bottrell, D.W., Suggested modification of pre-analytical holding times - volatile organics in water samples. *Proceedings of the Eleventh Annual Waste Testing & Quality Assurance Symposium,* 23-28 July, 1995, Washington Hilton Hotel and Towers, Washington DC; Bottrell, D.W., O.R. West, C.K. Bayne, R.L. Siegrist, and W.H. Holden, 1995. A confirmatory holding time study for purgeable VOCs in water samples. *Proceedings of the Twelfeth Annual Waste Testing & Quality Assurance Symposium,* 23-26 July, 1996, Washington Hilton Hotel and Towers, Washington DC.

change in results can be extended to 14 days.[8] However, even confronted with overwhelming scientific evidence, regulators and other legalists are frequently slow to move from their time-honored and codified positions. EPA Region IV, in a letter dated 19 October, 1999, is allowing soil samples to be mixed with water in a 40 mL VOA vial within 48 hours of collection, and then placed on their side in a freezer and frozen to give a holding time of 14 days for volatiles analysis.

This procedure may be acceptable within the general context of the Drinking Water regulations[9], which state that appropriate sample receipt/storage temperatures are 4 °C or below, as long as the glass vial does not break. Further, the wastewater quality assurance manual states,[10] "samples should be shipped and maintained at less than 4 °C."

The target value of 4 °C as a preservation objective is not defined in an analytically achievable sense. Measurement processes, be it temperature, length, or concentration measurements, have some degree of error inherent to the process. Acceptable margins of error are not associated with the value 4 °C. The U.S. Army Corps of Engineers is one of the few environmental regulatory agencies that have placed a measurement error upon 4 °C. The USACE acceptance range[11] is 2 °C to 6 °C, or a target value of 4 ± 2 °C.

Storage at a defined temperature is a different concept than sample receipt temperature. Samples initially received at the lab at an elevated temperature, say, for instance, 6 °C, are placed in the lab in temperature-controlled refrigerators that assure that the proper temperature is maintained for all samples. Thus the sample-received temperature is a transient event, most commonly occurring because the samples are still in the process of being cooled from the ambient temperature of the source of the samples.

The EPA Region IV sampling and shipping protocols[12] indicate that wastewater samples are to immediately be "iced" in a shipping container that contains 2-4 inches of vermiculite, ice in sealed bags or "blue ice" and then the rest of the space in the cooler filled with vermiculite. This is quite inefficient to cool samples as the vermiculite interferes with the heat transfer process. Our experience with this style of shipping indicates that receipt temperatures can vary from 6 °C to 15 °C, and rarely is 4 °C obtained. However, the EPA Region IV shipping procedure is by definition acceptable in meeting cooling requirements.

The U.S. Air Force Center for Environmental Excellences (AFCEE) has addressed these problems succinctly[13], "When a 4 °C requirement for preserving the sample is indicated, the samples shall be packed in ice or chemical refrigerant to keep them cool

---

[8] Turriff, D., C. Reitmeyer, L. Jacobs and N. Melberg, 1996. Performance of a new disposable sampling and storage device for soil VOCs. *Proceedings of the Twelfth Annual Waste Testing & Quality Assurance Symposium,* 23-26 July, 1996, Washington Hilton Hotel and Towers, Washington DC; Hewett, A, 1999. Freezer storage of soil samples containing volatile organic compounds. *Proceedings of the Fifteenth Annual Waste Testing and Quality Assurance Symposium*, 18-22 July, 1999, Crystal Gateway Marriott, Arlington, VA.

[9] USEPA, March, 1997. *Manual for the Certification of Laboratories Analyzing Drinking Water, Fourth Edition.* EPA 815-B-97-001, IV.6.2, page IV-3.

[10] USEPA, March, 1979. *Handbook for Analytical Quality Control in Water and Wastewater Laboratories.* EPA -600/4-79-019, page 8-2.

[11] USACE, July, 1994. *USACE Environmental Quality Engineering Manual: Validation of Analytical Chemistry Laboratories*, EM-200-1-1, page I-14.

[12] USEPA Region IV, May, 1996. *Environmental Investigations Standard Operating Procedures and Quality Assurance Manual*, Appendix D.

[13] AFCEE, March, 1998. *AFCEE Quality Assurance Project Plan, Version 3.0.* page 5-6 and 5-7.

during collection and transportation. During transit, it is not always possible to rigorously control the temperature of the samples. As a general rule, storage at low temperature is the best way to preserve most samples. While in the laboratory, samples shall be stored in limited-access, temperature-controlled areas. Refrigerators, coolers, and freezers shall be monitored for temperature seven days a week. Acceptance criteria for the temperature of the refrigerators and coolers is 4 °C ± 2 °C."

Defensible principles applied to evaluation of data then would suggest that samples held at temperatures less than 4 °C generate acceptable results. Samples held above 4 °C, but still analyzed within the adjusted holding time are acceptable. Samples held above 4 °C, and analyzed out of adjusted holding time need to have results qualified. The Bottrell paper, already cited as a reference, has extensive lists of practical retention times that are based on experiments.

The normal guidelines[14] for holding time non-compliance state that results from samples analyzed over holding times are to be considered as biased low. Thus a report of "non-detect" or "below detection limit" is to be considered a false negative, in that the analyte may originally have been in the sample, but has degraded away. Analytes reported as present are quantified lower than they actually are in the source, thus the values are considered low estimates.

A storage temperature vs. holding time scale can be generated using the Arrhenius equation to define the slope for a test with a nominal 28-day holding time (Table 9-1). This assumes that there is a regular relationship between the temperature and the factors that cause degradation of the target analytes. Unfortunately biological degradation exhibits a more complex relationship that includes population growth/decline and temperature-dependent chemical metabolism, commonly a discontinuous function complicated by coma/death.

Table 9-1. Theoretical dependence of storage temperature on holding time (nominally 28 days at 4 °C).

Storage Temperature (°C)	Maximum holding time (days)
0	39
2	33
4	28
6	25
8	22
10	20
12	17
14	14

[14] Berger, W., H. McCarty, and R.-K. Smith, 1996. *Environmental Laboratory Data Evaluation*, Genium Publishing, Schenectady, NY; USEPA, October, 1999. *National Functional Guidelines for Organic Data Review*, EPA-540/R-99/008, www.epa.gov/oerrpage/superfund/programs/clp/download/fgorg.pdf; USEPA, February 199, *National Functional Guidelines for Inorganic Data Review,* EPA-540/R-94-013; USEPA, June 1993, *Guidance on Evaluation, Resolution, and Documentation of Analytical Problems Associated with Compliance Monitoring*, EPA 821-B-93-001.

B. Laboratory Documentation

Many people develop bad habits during high school and college laboratory classes that are not acceptable in the environmental community. These proscribed habits arise from a cavalier attitude to documentation, ranging from writing weights on their hand to rote following of a written procedure without taking any notes on observations. The emphasis upon detailed record keeping is a direct result of governmental enforcement actions of the laws, rules, and regulations that form the framework for the environmental business and laboratory operations in general. Enforcement leads directly to involvement with the courts system, where laboratory records frequently are essential evidence for either or both sides of the case. The Federal Rules of Evidence and their application to laboratory records have been described elsewhere in detail.[15]

The starting place for examining a laboratory's records is the Quality Assurance Manual.[16] This should be a complete description of all the activities of the laboratory quality division. The library should have a complete collection of the Methods that are being used.

From the Methods, a Standard Operating Procedure[17] (SOP) is prepared. To be useful an SOP must be more than a photocopy of the Method. Methods are commonly written in very broad terminology such as, "calibrate the instrument at five concentration levels ranging from the lowest reporting limit to the highest reportable value." The SOP serves to fill in the gaps in these directions, for example, a detailed description of where to purchase the standards, how to prepare them, and at exactly what concentrations they should be. The SOP should also describe in detail how the instrument software is manipulated to achieve the calibration and sample quantitation. The idea is to prepare a supplement to the Method that allows a new analyst to successfully perform the procedure with minimal direction from an experienced person.

The Quality Assurance Manual, Method, and SOP serve to provide a framework of operations that allow analysis, but do not by themselves provide evidence of the fact of the sample analysis. These facts are established through the supporting records of the laboratory, which should include standard and reagent preparation logs, sample preparation logs, calibration and calibration verification records, instrument run logs, chromatograms and quantitation reports for samples and quality control measures, method detection limit studies, initial demonstration studies, sample receipt, storage and disposition records, and performance evaluation results. The visitor to the laboratory can check these records by choosing a sample at random, then tracing the analysis of the sample through the on-hand records.

The bottom line with documentation is that there must be a paper trail that can recreate the complete event of the analysis. Further, it should be documented in

---

[15] Berger, W, H. McCarty, and R.-K. Smith, 1996. *Environmental Laboratory Data Evaluation*, Genium Publishing, Schenectady, NY; Giannelli, P.C., and E.J. Imwinkelried, 1993. *Scientific Evidence*, Second Edition, Michie Company, Charlottesville, VA; Graham, M.H., 1992. *Federal Rules of Evidence*, Third Edition, West Publishing, St. Paul, Minn; Imwinkelried, E.J., 1997. *The Methods of Attacking Scientific Evidence*, Third Edition. Lexis Law Publishing, Charlottesville, VA ISBN 1-55834-777-1, 1-800-446-3410

[16] USEPA, ORD, 1997. *Guidance for Quality Assurance Project Plans.* EPA QA/G-5 Available on the Internet at www.epa.gov/ord/qa; NELAP Standards, 1999, 5. Quality Systems, www.epa.gov/ttn/nelac

[17] USEPA, ORD, 1995. *Guidance for the Preparation of Standard Operating Procedures (SOPs) for Quality-Related Documents.* EPA QA/G-6, EPA/600/R-96/027. Available on Internet at www.epa.gov/ord/qa; NELAP Standards, 1999, 5. Quality Systems, www.epa.gov/ttn/nelac

sufficient clarity that the paper trail can be collected without the presence of the person who originally created it.

C. Performance Evaluation Samples

Performance evaluation (PE) samples test the ability of the analyst to determine the correct concentration of target analytes in an unknown sample. In most PE samples the identification of all the analytes is known, and the task is accurate quantitation. In other samples the PE sample contains a subset of a list of possible target analytes. The analyst must correctly identify and quantitate each of the analytes present to obtain an acceptable evaluation. These are sometimes called challenge samples and are frequently encountered in organic analysis.

Historically there have been two major sets of PE samples performed by most environmental laboratories. These are the Water Supply (WS) and Water Pollution (WP) studies, provided by the EPA at no cost to the participants. The samples were sent to labs twice a year as either liquid concentrates in sealed glass ampules or as solids. In either case the sample had to be diluted to a prescribed volume and then analyzed. There are significant differences between the manufacture and evaluation of the two types of studies.

The WS studies were - and still are - an integral part of the drinking (potable) water laboratory certification program. Evaluation guidelines for the WS studies exist in the Federal Regulations at 40 CFR 141. Results graded as acceptable had to be within a set percentage of the true (weighted out) value of the analyte in the sample. The acceptance ranges for the organic analytes are presented in Table 9-2.

Table 9-2. Acceptance ranges for regulated organic analytes in WS studies (from 40 CFR 141).

Analyte	Acceptance range
Total Trihalomethanes	±20%
Regulated VOC	±20% if >10 ug/L
	±40% if <10 ug/L
Alachlor	±45%
Atrazine	±45%
Carbofuran	±45%
Chlordane	±45%
2,4-D	±50%
Dalapon	2 Std. Deviations[18]
Dibromochloropropane (DBCP)	±40%
Dinoseb	2 Std. Deviations
Diquat	2 Std. Deviations
Endothall	2 Std. Deviations
Endrin	±30%
Ethylenedibromide (EDB)	±40%
Glyphosate	2 Std. Deviations
Heptachlor	±45%

[18] Two standard deviations based on the results of the reference laboratories used in the QA Study. The reference labs are the EPA Regional labs and selected Primacy State labs.

Table 9-2. Acceptance ranges for regulated organic analytes in WS studies (from 40 CFR 141), *continued*

Analyte	Acceptance range
Heptachlor epoxide	±45%
Lindane	±45%
Methoxychlor	±45%
Oxamyl	2 Std. Deviations
Pentachlorophenol	±50%
Picloram	2 Std. Deviations
Simazine	2 Std. Deviations
Toxaphene	±45%
2,4,5-TP (Silvex)	±50%
Hexachlorobenzene	2 Std. Deviations
Hexachlorocyclopentadiene	2 Std. Deviations
Benzo(a)pyrene	2 Std. Deviations
PCB (as DCB)	0-200%
Di(2-ethylhexyl)adipate	2 Std. Deviations
Di(2-ethylhexyl)phthalate	2 Std. Deviations

EPA would evaluate the results from the participants in the study statistically, then compare them to the regulatory acceptance range. In cases where a mistake occurred in the preparation of the PE sample as evidenced by the mean of the results deviating significantly from the weighed-out value of the analyte, the only recourse for EPA was to invalidate the affected analytes for that study, rather than simply adjust the acceptance range.

The absolute regulatory acceptance ranges sometimes create real problems for the participants in the WS studies. If the concentration of the analyte in a sample is very low, a very small acceptance range exists. For instance if the analyte's true value is 2 µg/L with an acceptance range of ± 50%, then acceptable values must fall in the range 1-3. On the other hand, a value of 10 µg/L generates an acceptance range of 5-15 µg/L. This is the opposite to what is normally observed in organic analysis, where very low levels of analyte correlate to greater absolute variation in results.

The WP study acceptance ranges are generated by taking the results from the participants and calculating the mean and standard deviation (sd). The warning limits for the study are the ± 2 sd interval around the mean, while the acceptance range is the ± 3 sd interval.

To summarize then, the acceptance ranges for the WS studies are pre-set based on the weighed-out true value of the analyte in the sample, while the WP studies are evaluated based on the participant results. The latter procedure is the only scientifically rational evaluation. However, regulators are extremely reluctant to delete existing regulations. If changes are needed, then an additional regulation is prepared and promulgated, which, in essence, says that the regulated community does not have to comply with the earlier regulation.

The WS and the WP programs are not the only PE samples that EPA provides. Other programs are the Discharge Monitoring Report Quality Assurance studies (DMR-QA), and specialized PE studies to support assorted programs. One example was the disinfection by-product information collection study (DBP-ICR), which lasted from 1993 to 1999. Participant laboratories had to successfully complete DBP PE studies.

In 1999, the EPA, in a budget-cutting measure, ceased to provide the WS and WP studies. The requirement for participation in the studies still exists: however, laboratories have to obtain the samples from approved commercial suppliers. The EPA sets the parameters and provides oversight for the replacement WS and WP studies, and the National Institute for Standards and Technology (NIST) provides the accreditation of the suppliers. For current information about the list of approved suppliers, the National Environmental Laboratory Accreditation Program (NELAP) website should be accessed (www.epa.gov/ttn/nelac).

PE study results are an essential element in the evaluation of laboratories. The commercial suppliers of the samples do not limit themselves to just the analytes and matrices of the WS and WP studies. There are PE samples available for every type of analyte in a wide variety of matrices such as soils, air, sludges, and realistic wastewater effluents. Several suppliers will prepare custom PE samples with material collected on project sites. These prepared samples can be added into the flow of samples being sent to the laboratory as a blind check on the analytical performance.

The interpretation of the results from the WS and WP is fairly straightforward, and normal performance by a laboratory should be to get at least 85% of the results within the acceptance range. Values outside the acceptance range should be checked very closely to find out how close the reported value is to bounds of the acceptance range. The laboratory quality assurance manager or technical director should investigate each unacceptable result and prepare a written response explaining the incorrect results. This response should always be attached to the PE study evaluation and should be checked by any subsequent evaluator.

A common finding is that the reported value for an unacceptable analyte is just barely outside the range. In other cases there are substantial differences between the acceptance range and the lab's answer. Most often there are simple explanations for these deviant results.

Sometimes it is just a matter of how the laboratory rounded the result. Many acceptance ranges are listed with 3 or 4 significant figures, regardless of any scientifically achievable sensitivity. For example, for a volatiles analysis by EPA Method 624 the acceptance range might be calculated as 2.34 - 5.68 µg/L. There is no commercial laboratory in the US that can report a reproducible volatiles result to the sensitivity of 0.01 µg/L, and a reputable lab will round a result of 5.68 to 5.7 µg/L. The scientifically defensible result of 5.7 µg/L is interpreted by the computer that is performing the evaluation as 5.70 µg/L which is outside the acceptance range, and an evaluation of unacceptable will be generated. These problems are more frequently seen when analytes are in the lower concentration ranges, but can occur with any reported result.

Another common source of unacceptable results is due to the necessity to dilute the concentrate or solid to a specified volume. Mistakes occur in the dilution process when different samples have different dilution directions. A clue that this may be the case is when the analyte result is a factor of 5, 10 or 20 outside the acceptance range. An analyst who assumes that the directions for the in-hand sample are the same as they were for the last PE sample will be prone to making these errors. The directions must be read carefully for each sample. The dilution requirement also causes problems because some samples must be reported based on the concentration of the analytes in the dilution, while others are reported based on the concentrate. Again these common errors can be avoided by carefully reading the directions for each and every sample. An evaluator of the PE sample results must keep in mind that these extra manipulations and calculations are outside the bounds of normal sample analysis. While they indicate that the analysts may not be paying careful attention to what they are doing, they have little bearing on the ability to handle real samples.

When a laboratory consistently scores above 95% and, particularly, when a lab gets 100% on PE studies this it is a cause for caution rather than any reason for celebration. It is a distinct sign that the lab is spending an inordinate amount of time analyzing, re-analyzing, and performing non-routine techniques on the PE sample to insure that they get the right answer. I particularly like to use a multiple level standard addition technique on PE samples to obtain the absolutely best answer possible. But these manipulations are well beyond the attention that a normal sample receives and are not representative of anything other than the cleverness of the analyst in tackling the PE sample. The time spent on assuring that the exactly correct answers are obtained on a PE study is time that is non-revenue producing. Commercial labs exist based on having a positive accounting sheet, and, when overhead is increased due to PE samples, the overhead must be reduced elsewhere. The most common area is in less time and effort spent on revenue-producing samples. Labs that are treating the PE samples about like they treat real samples will have overall scores that fluctuate. Sometimes overall scores as low as 80-85% will be obtained, and every once in a while a near perfect score will be logged. But the variation in test scores is a reassuring reflection of normal operational randomness.

Evaluators should take extra care when examining results on PE samples prepared from site-specific materials. These are generally analyzed by only one or two labs. Direct comparisons of results between the labs and the cited true value must be done very cautiously. How the PE sample preparer determined the true value of the analytes is a significant area of concern. The weight of the analyte added to the sample, particularly when the sample is a soil or sludge, is only a rough guide to how much analyte is recoverable from the sample. Recoverable is the key concept. Different extraction techniques will have different efficiencies. A sonication extraction (EPA Method 3550) with methylene chloride has a different efficiency for any specific sample than sonication with a mixed acetone-methylene chloride solvent. So the PE sample preparer must verify the recoverability of the analytes using techniques that are exactly the same as those that the participant labs will use. This concept also applies when comparing the results of the tested labs. Apples can not be compared to oranges, and even apples have many varieties. A granny smith is different than a golden delicious although, they are both apple varieties.

If significant variations are noted between the labs, this may be an indication that different techniques are being used. The solution may be to change the sample processing techniques to the more efficient techniques. Decisions to make these changes must be based on observation of consistent differences. Basing a process change on only the result of a single PE sample is very risky, as it may well be that the PE sample is the exception rather than being representative of the normal sample behavior. Remember that the PE sample was processed by the PE sample preparer in a significantly different manner than the normal site sample would receive when analyzed by the lab.

CHAPTER 10

References

1. AFCEE, March, 1998. *AFCEE Quality Assurance Project Plan, Version 3.0.*

2. Berger, W., H. McCarty, and R.-K. Smith, 1996. *Environmental Laboratory Data Evaluation.* Genium Publishing, Schenectady, NY 1-800-243-6486.

3. Bottrell, D.W., 1995. Suggested modification of pre-analytical holding times - volatile organics in water samples. *Proceedings of the Eleventh Annual Waste Testing & Quality Assurance Symposium,* 23-28 July, 1995, Washington Hilton Hotel and Towers, Washington DC.

4. Bottrell, D.W., O.R. West, C.K. Bayne, R.L. Siegrist, and W.H. Holden, 1995. A confirmatory holding time study for purgeable VOCs in water samples. *Proceedings of the Twelfth Annual Waste Testing & Quality Assurance Symposium,* 23-26 July, 1996, Washington Hilton Hotel and Towers, Washington DC.

5. Buffington, R., and M.K. Wilson, 1991. Detectors for Gas Chromatography - A Practical Primer. Hewlett-Packard, Co., Wilmington, DE.

6. Castellan, G.W.,1971. *Physical Chemistry, Second Edition*, Addison-Wesley Publishing Company, Reading, MA.

7. Dixon, W.J., 1953. Processing data outliers, *Biometrics* 9(1):74-89.

8. Giannelli, P.C., and E.J. Imwinkelried, 1993. *Scientific Evidence*, Second Edition, Michie Company, Charlottesville, VA.

9. Glassmeyer, S.T., K.E. Shanks, and R.A. Hites, 1999. Automated toxaphene quantitation by GC/MS. *Anal. Chem.* 71(7):1448-1453.

10. Graham, M.H., 1992. *Federal Rules of Evidence*, Third Edition, West Publishing, St. Paul, Minn.

11. Grubbs, F.E., and G. Beck, 1972. Extension of sample sizes and percentage points for significance tests of outlying observations. *Technometrics* 14(4):847-854.

12. Hargis, L.G., 1988. *Analytical Chemistry Principles and Techniques*, Prentice-Hall, Englewood Cliffs, NJ

13. Heller, J., 1961. *Catch 22*. Simon & Schuster, New York, NY.

14. Hewett, A,1995. Determining volatile organic compound concentration stability in soil. *Proceedings of the Eleventh Annual Waste Testing & Quality Assurance Symposium,* 23-28 July, 1995, Washington Hilton Hotel and Towers, Washington DC.

15. Hewett, A, 1999. Freezer storage of soil samples containing volatile organic compounds. *Proceedings of the Fifteenth Annual Waste Testing and Quality Assurance Symposium*, 18-22 July, 1999, Crystal Gateway Marriott, Arlington, VA.

16. Hewlett-Packard, Inc., 1996. HP 6890 Series Gas Chromatograph Operating Manual. Hewlett-Packard, Co., Wilmington, DE.

17. Hites, R.A., 1992. *Handbook of Mass Spectra of Environmental Contaminants, Second Edition.* Lewis Publishers, Boca Raton, FL.

18. Hyver, K.J., and P. Sandra, 1989. High Resolution Gas Chromatography, Third Edition. Hewlett-Packard, Co., Wilmington, DE.

19. Imwinkelried, E.J., 1997. *The Methods of Attacking Scientific Evidence*, Third Edition. Lexis Law Publishing, Charlottesville, VA ISBN 1-55834-777-1, 1-800-446-3410.

20. Jackson, C., 1996. Analysis of phenolic acid compounds in calcareous soils by SW-846 Method 8270, *Proceedings of the 19th Annual EPA Conference on Analysis of Pollutants in the Environment,* Norfolk VA 15-16 May, 1996.

21. Kannan, K., T. Imagawa, A.L. Blankenship, and J.P. Giesy, 1998. Isomer-specific analysis and toxic evaluation of polychlorinated naphthalenes in soil, sediment, and biota collected near the site of a chlor-alkali plant. *Environ. Sci. Technol.* 32(17):2507-2514.

22. Kishi, H., H. Arimoto, and T. Fujii, 1998. Analysis of alcohols and phenols with a newly designed gas chromatographic detector. *Anal. Chem.* 70(16):3488-3492.

23. Lough, W.J., and I.W. Wainer, 1992. *High Performance Liquid Chromatography, Fundamental Principles and Practice*. Blackie Academic & Professional, New York, NY.

24. Marsh, G., J. Hu, E. Jakobsson, S. Rahm, and A. Bergman, 1999. Synthesis and characterization of 32 Polybrominated diphenyl ethers. *Environ. Sci. Technol.* 33(17):3033-3037.

25. Massachusetts Department of Environmental Protection. EPH (extracted petroleum hydrocarbon www.state.ma.us/dep/bwsc/vph_eph.htm)

26. McLafferty, F.W., and F. Turecek, 1993. *Interpretation of Mass Spectra, Fourth Edition.* University Science Books, Sausalito, CA.

27. Meyer, V.R., 1994. Practical High-Performance Liquid Chromatography, Second Edition. John Wiley & Sons, New York, NY.

28. NELAP Standards, 1999, Section 5. Quality Systems, www.epa.gov/ttn/nelac

29. Persson,B.-A., J. Vessman, and R.D. McDowell, 1998. Is your method specific or just selective. *LC•GC*, 16(6):556-560, June, 1998.

30. Rosner, B., 1975. On the detection of many outliers. *Technometrics* 17:221-227.

31. Silverstein, R.M., G.C. Bassler, and T.C. Morrill, 1991. *Spectrophotometric Identification of Organic Compounds, Fifth Edition*. John Wiley & Sons, New York, NY.

32. Smith, R.-K., 1990. Chemotaxonomy of honey bees (*Apis mellifera* L.). Part 1: European and African workers. *Bee Science* 1(1):23-32

33. Smith, R.-K., 1991. Chemotaxonomy of honey bees (*Apis mellifera* L.). Part 2: Africanized workers. *Bee Science* 1(2):82-94

34. Smith, R.-K., 1997. Discussion of: Organic Priority Pollutants in New York City Municipal Wastewaters: 1989-1993, A.I. Stubin, T.M. Brosnan, K.D. Porter, L. Jimenez and H. Lochan, *Water Environ. Res.* 1997 69(3):382 May/June 1997 issue.

35. Smith, R.-K., 1998. Benzidine? Really? *Proceedings of the 14th Annual Waste Testing and Quality Assurance Symposium*, 13-15 July, 1998, Crystal Gateway Marriott, Arlington VA., pp. 40-43.

36. Smith, R.-K., 1999. *Lectures on Wastewater Analysis and Interpretation.* Genium Publishing, Schenectady, NY.

37. Smith, R.-K., 1999. *Handbook of Environmental Analysis, Fourth Edition.* Genium Publishing, Schenectady, NY.

38. Smith, R.-K., M. Spivak, O.R. Taylor, Jr. , C. Bennett, and M.L. Smith, 1992. Chemotaxonomy of honey bees (*Apis mellifera* L.). Part 3: Identification of Africanization in honey bee queens (Hymenoptera: Apidae). *Bee Science* 2(2): 93-105

39. Solsky, J., 1998. Questionable practices in the Organic Laboratory. *Proceedings of the 14th Annual Waste Testing and Quality Assurance Symposium*, 13-15 July, 1998, Crystal Gateway Marriott, Arlington VA.

40. Solsky, J., 1999. Questionable practices in the Organic Laboratory: Part II. *Proceedings of the 15th Annual Waste Testing and Quality Assurance Symposium*, 18-22 July, 1999, Crystal Gateway Marriott, Arlington VA.

41. Taylor, J.K.,1987. *Quality Assurance of Chemical Measurements*, Lewis Publishers, Chelsea MI

42. Tomy, G.T., J.B. Westmore, G.A Stern, D.C.G. Muir, and A.T. Fisk, 1999. Interlaboratory study on quantitative methods of analysis of C10-C13 polychloro-n-alkanes. *Anal Chem.* 71(2):446-451.

43. Turriff, D., R. Reitmeyer, L. Jacobs, and N. Melberg, 1995. Comparison of alternatives for sampling and storage of volatiles in soil. *Proceedings of the Eleventh Annual Waste Testing & Quality Assurance Symposium,* 23-28 July, 1995, Washington Hilton Hotel and Towers, Washington DC.

44. Turriff, D., C. Reitmeyer, L. Jacobs and N. Melberg, 1996. Performance of a new disposable sampling and storage device for soil VOCs. *Proceedings of the Twelfth Annual Waste Testing & Quality Assurance Symposium,* 23-26 July, 1996, Washington Hilton Hotel and Towers, Washington DC.

45. U.S. Army Corps of Engineers, October, 1998. Shell for Analytical Chemistry Requirements, Omaha, NE.

46. U.S. Army Corps of Engineers, July, 1994. *USACE Environmental Quality Engineering Manual: Validation of Analytical Chemistry Laboratories*, EM-200-1-1

47. USEPA, 1995. *Analytical Methods for the Determination of Pollutants in Pharmaceutical Manufacturing Industry Wastewater* (EPA-821-94-001) February, 1995

48. USEPA, 1979. *Handbook for Analytical Quality Control in Water and Wastewater*, EPA-600/4-79-019, NTIS PB-297451

49. USEPA, April, 1994. *Industrial User Inspection and Sampling Manual for POTW's*, EPA 831-B-94-001

50. USEPA, October, 1999. *National Functional Guidelines for Organic Data Review*, EPA-540/R-99/008, www.epa.gov/oerrpage/superfund/programs/clp/download/fgorg.pdf

51. USEPA, February 1994, *National Functional Guidelines for Inorganic Data Review,* EPA-540/R-94-013.

52. USEPA, June 1993, *Guidance on Evaluation, Resolution, and Documentation of Analytical Problems Associated with Compliance Monitoring*, EPA 821-B-93-001.

53. USEPA, ORD, 1995. *Guidance for the Preparation of Standard Operating Procedures (SOPs) for Quality-Related Documents.* EPA QA/G-6, EPA/600/R-96/027. Available on Internet at www.epa.gov/ord/qa.

54. USEPA, ORD, 1997. *Guidance for Quality Assurance Project Plans.* EPA QA/G-5 Available on the Internet at www.epa.gov/ord/qa.

55. USEPA, ORD, 1998. *Guidance for Data Quality Assessment: Practical Methods for Data Analysis.* EPA QA/G-9, EPA 600/R-96/084, January, 1998. Available on the Internet at www.epa.gov/ord/qa.

56. USEPA, March, 1997. *Manual for the Certification of Laboratories Analyzing Drinking Water, Fourth Edition*, EPA 815-B-97-001, available on the Internet from the EPA website, www.epa.gov/ogwdw

57. USEPA. *Test Methods for Evaluating Solid Waste - Physical/Chemical Methods,* EPA/SW-846, 3rd Edition, 1986, Update 1, July, 1992, Updates II and IIa, 1994, Update III, 1996, and proposed Updates IVa and IVb, 1998. Available from the Internet at www.epa.gov/osw/methods or by subscription from the Government Printing Office.

58. USEPA Region IV, May, 1996. *Environmental Investigations Standard Operating Procedures and Quality Assurance Manual.*

59. Walsh, J.E. 1958. Large sample nonparametric rejection of outlying observations. *Annals of the Institute of Statistical Mathematics*, 10:223-232.

60. Watson, J.T., 1997. *Introduction to Mass Spectrometry, Third Edition.* Lippincott-Raven, Philadelphia, PA.